EXPERIMENTS FOR EXPLORING ELECTRICITY

Howard B. Brown
Massachusetts Bay Community College

Prentice Hall
Englewood Cliffs, New Jersey   Columbus, Ohio

© 1996 by Prentice-Hall, Inc.
A Simon & Schuster Company
Englewood Cliffs, New Jersey 07632

All rights reserved. No part of this book may be
reproduced in any form or by any means,
without permission in writing from the publisher.

Printed in the United States of America

10 9 8 7 6 5 4 3 2 1

ISBN 0-13-207580-6

Prentice-Hall International (UK) Limited, *London*
Prentice-Hall of Australia Pty. Limited, *Sydney*
Prentice-Hall Canada Inc., *Toronto*
Prentice-Hall Hispanoamericana, S.A., *Mexico*
Prentice-Hall of India Private Limited, *New Delhi*
Prentice-Hall of Japan, Inc., *Tokyo*
Simon & Schuster Asia Pte. Ltd., *Singapore*
Editora Prentice-Hall do Brasil, Ltda., *Rio de Janeiro*

Dedication

To: My wife and children,

   . . . for their special help in creating this book.

Preface

This manual is a series of electronic laboratory projects designed to be an integral part of a first course in DC/AC electronics. The author's goal is to create a lab manual that is easier for both instructor and student to use—a book that generates motivation, interest and a sense of learning for the student. In summary, this is a book that is *user friendly* for both students and teachers.

Interactive Presentation

A key element in the manual's format is the interactive style of presentation. Projects are designed to involve students with the results of their activities, provide immediate feedback, and indicate success or the need for review.

Interactive Procedures: Typical instructions ask the student to think about a result, make a prediction, and commit to an answer *before* performing the activity. This personal challenge and small amount of *risk* involves the learner more completely. It contrasts with a set of instructions which may be treated by the student as simply a job to be done or numbers to be obtained. After performing the task, the student compares results with prediction.

Experimental Errors: The concept of measurement deviation or laboratory error is introduced early. Students are given a method to calculate the error and a limit value. If their data are within these limits, they know their results are good. This generates a sense of accomplishment and personal satisfaction. If the error is too large, the student knows the project needs to be checked. They quickly learn to do this on their own and obviate much of the need to call the instructor for approval.

Guided Discovery: Data sets are analyzed to observe or discover points of theory discussed in class. (Experiment 21, Maximum Power Transfer Theorem is an example of this type of project.) Individual experimental results are also used to reinforce elements of theory.

Graphical Analysis: Emphasis is placed on drawing and interpreting graphs. Data are plotted where appropriate, and graphical analysis is used to observe or discover points of theory. Examples of projects include linear versus nonlinear resistance and resonant frequency response.

Style and Format

Data Entry: Measurements are entered on the same instruction line that describes the activity to be performed. Students do not need to search through pages to find the location for marking answers. A data table may also be provided for summary, compilation, and comparison.

Procedures: The instructions are written to lead students through activities with minimum input from the instructor. The self-paced and sometimes conversational style helps create a comfortable flow of activity. The student is "walked through" the procedures at a pace which the author's experience has shown to be effective.

Contents

Variety of Components: The manual includes projects for several common, inexpensive components. These include the potentiometer in its three functions, electromagnetic relay, thermistor, the photoresistor, incandescent lamp, and wire measurements making use of the AWG wire tables.

Component Assembler Boards: The manual contains two projects specifically designed to assist students in learning the connection patterns of the common PC breadboard.

Troubleshooting: Problem solving is built into several topics. Students analyze a properly working circuit and then create a problem (open, short, or wrong value) to observe the effects of this problem on circuit operation.

Learning Through Drill: Projects typically require several repetitions of circuit calculations and measurements, with different values for each variable.

Lab Questions: A series of questions based on the current lab results is included at the end of each project. Students may be asked to analyze their observations or data and sometimes perform an additional short procedure to confirm their answers. Typically, these are not questions that can be answered from theory alone, but require the current activity combined with theoretical principles for a solution.

Pedagogy

The manual employs several techniques to enhance the student's learning and motivation. Student involvement, interactive instructions, the use of experimental error for immediate feedback, guided discovery, and repetition or drill—all contribute to the educational effectiveness.

Contents

Experiment	DC Circuits	
1	Resistance Measurement and the Resistor Color Code	1
2	Voltage Measurement and Experimental Error	5
3	Current Measurement and Laboratory Procedure	9
4	Ohm's Law	13
5	Graphing the Volt-Ampere Characteristic for a Nonlinear Resistance	17
6	Resistance Measurement Applications	25
7	Breadboard Connections	31
8	Series Circuits 1: Normal Operation	35
9	Series Circuits 2: The Effect of Opens and Shorts	41
10	The Voltage Divider: Operation and Calculation	45
11	Parallel Circuits 1: Normal Operation	51
12	Parallel Circuits 2: The Effect of Opens and Shorts	57
13	Series-Parallel Circuits 1: Normal Operation	61
14	Series-Parallel Circuits 2: The Effect of Opens and Shorts	67
15	Voltmeter and Circuit Loading	73
16	The Variable Resistor	79
17	The Potentiometer and Voltage Control	81
18	The Rheostat and Current Control	83
19	Relays and Relay Circuits	87
20	Internal Resistance and the Loaded Power Source	93
21	Maximum Power Transfer Theorem	97
22	Thevenin's Theorem	103
23	The Wheatstone Bridge	109

Experiment	AC Circuits	
24	Oscilloscope and Waveform Generator: Familiarization	113
25	Measuring Time and Frequency with the Oscilloscope	121
26	Measuring AC Voltage with the Oscilloscope	127
27	Transformer Fundamentals	131
28	*RC* Time Constants	135
29	Capacitance and Capacitive Reactance	141
30	Series and Parallel Capacitance	145
31	Series *RC* Circuits	149
32	Phase Measurements	155
33	Parallel *RC* Circuits	161
34	Inductance and Inductive Reactance	167
35	Series *RL* Circuits	173
36	Parallel *RL* Circuits	179
37	Series *RLC* Circuits	183
38	Series Resonance	189
39	Parallel *RLC* Circuits	201
40	*RC* Filters	208

1 Resistance Measurement and the Resistor Color Code

Objectives

1. Develop skill in reading the resistor color code.
2. Use and read the analog or digital ohmmeter.
3. Apply the concept of "tolerance" to component measurement.

Preparation

As with all lab projects, you need to study the subject *before* performing the laboratory exercises. For this project, you should be familiar with the EIA color code, the calculation of resistor tolerance, and the proper use of your laboratory ohmmeter.

You will need a set of 10 color-coded resistors. A selection of appropriate values is included in Table 1-1.

Materials

One VOM meter
Resistors:
A set of 10 resistors, 1/2 W or larger. The values are optional, but a selection from each decade from 100 Ω to 1 MΩ is suggested. A list of optional resistance values is given in Table 1-1.

Table 1-1 Suggested Resistor Values

100 Ω	10 kΩ
330 Ω	22 kΩ
470 Ω	39 kΩ
1 kΩ	56 kΩ
1.5 kΩ	100 kΩ
3.9 kΩ	270 kΩ
5.6 kΩ	470 kΩ
8.2 kΩ	1 MΩ

Procedure

Steps 1–8 below refer to the numbered columns of Table 1-2. The headings of the table give a brief description of the content of the columns or the procedure to be followed. Steps 9 and 10 give instructions for completing and evaluating the results.

1. Select one resistor from the set and read the color code resistance.

2. Enter the color code resistance in ohms in column 2.

3. Convert the ohms value to kilohms; enter in column 3.

4. Determine the resistor tolerance from the color code and enter in column 4.

5. Next, using the ohmmeter, measure the resistor and record your measurement in column 5.

6. Calculate the high resistance limit and enter in column 6.

7. Calculate the low resistance limit and enter in column 7.

8. In column 8, examine your results. Is the measured value of resistance between the high and low limits for this resistor? If it is, it indicates a good measurement and reading of the color code.

    If your measured value is not between the high and low limits, it may indicate a poor resistor. It may also indicate an incorrect reading of the meter or the color code.

9. Complete the data for the remaining nine resistors.

10. After completing the table, look at your results in column 8.

    If your measurements are all *within* the tolerance limits, you can feel confident in your ability to read the color code and use the ohmmeter.

    If most values are *not within* tolerance, you need to check your measurements, color code readings, or the proper use of the ohmmeter.

Ohm's Law $I = V/R$

2 Voltage Measurement and Experimental Error

Objectives

1. Develop skill in reading the voltmeter.
2. Determine the experimental error or percent difference between measured and calculated values.
3. Apply the concept of experimental error to laboratory results.

Preparation

The circuit consists of two resistors connected in series. You will take a voltage reading across each resistor and add the two values. If the sum equals the supply voltage, you know your readings are *correct*. If the two voltages do *not* add up to the source value, it indicates an *incorrect* reading and you need to check the meters, circuit connections, or ask instructor for assistance.

Now, the questions arise: What if the numbers are not *exactly* the same? What if they are close? How different can the numbers be and still be acceptable?

Experimenters in all sciences must deal with the fact that measurements are *not* exact. Errors occur due to equipment accuracy, readability of dials, interpretation of meter display, rounding of decimal places, or the *error gremlin* that happens to pick *your* bench on a particular day.

There is an easy way to determine if your measurement results are close enough to indicate that your readings and calculations are *probably* correct. A tolerance value for the experimental difference between measurement and calculation is selected by the instructor or indicated in the lab manual. A suggested tolerance for DC projects is +/-5%.

If the difference between your measured and calculated values is less than this, you can feel confident that your results are good. If the difference between measured and calculated values is *too large*, it indicates you need to check the reading, circuit wiring, or component values.

The formula to calculate the percent error or difference is:

$$\% \text{ difference} = \frac{\text{measured} - \text{calculated}}{\text{calculated}} \times 100$$

Example: A student *calculates* voltage as 4.5 V but *measures* 4.6 V. The percent difference for these two values is:

$$\% = \frac{4.6 \text{ V} - 4.5 \text{ V}}{4.5 \text{ V}} \times 100 = 2.2\%$$

Result: 2.2% is an acceptable difference, but it is still possible that a more careful reading or experimental technique would yield even closer values.

Example: A student *calculates* resistance as 600 Ω but *measures* 450 Ω. The percent difference for these two values is:

$$\% = \frac{450 \text{ Ω} - 600 \text{ Ω}}{600 \text{ Ω}} \times 100 = -25\%$$

Result: −25% is much more than the allowed tolerance of 5%; therefore, the student needs to do some checking to find the cause of the error. Usually, when the problem has been found, a much smaller difference will result.

Note: The *minus sign* indicates the measured value is less than the calculated value.

Materials

DC power supply
One VOM meter
Resistors: 1/2 watt
   One 2.2 kΩ, one 5.6 kΩ

Procedure

1. Build the circuit shown in Figure 2-1. Turn the voltage source to zero and set the voltmeter to the 10 V range.

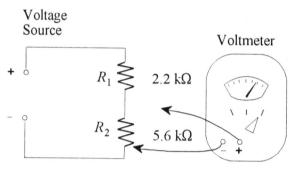

Figure 2-1

2. Slowly increase the source voltage to 2 V. Take readings of V_1 and V_2 and record the data in Table 2-1.

3. Add the two voltage readings and enter the sum in column 4. Theoretically this value should equal the total applied voltage.

4. Compute the percent difference between the voltages in column 4 and column 1. Enter this in column 5.

5. Increase the supply voltage to the next value indicated. Continue to take readings and calculate the experimental difference for each line.

Note: It is important to complete each line of data horizontally.

It would be a mistake in procedure to take readings *down* each column since you would have no indication that your data were correct until the project was completed. If you had some major errors in readings, you would need to repeat the entire project.

If the percent difference is calculated on each line, you will know right away if the reading is correct, or reasonably close.

6 Resistance Measurement Applications

Objectives

1. Use the American Wire Gauge (AWG) wire table to find the length of a spool of wire.
2. Build a photoresistor circuit to indicate relative light intensity.
3. Build a thermistor circuit to indicate changes in temperature.

Preparation

Part 1—AWG Wire Table: You will need several spools of wire of different lengths and sizes. Each spool should have access to the inner end of the wire and a label or specification that states the AWG wire size. You will also need to use the wire tables from your textbook.

Several factors determine the resistance of the wire: length, diameter, material, and temperature. If the spool of wire you are using is of the same material and at the same temperature for which the table was written, you can find the length of the wire easily. The procedure demonstrates the method for finding wire length.

Part 2—Photoconductive Cell or Photoresistor: A photoresistor is a semiconductor device whose resistance changes with light intensity.

The procedure calls for variations in light from complete darkness to bright light. You'll need to use some ingenuity to find the variation in light intensity required.

Part 3—Thermistor: There are many types of thermistors available, and each has its own table of temperature-resistance conversion. The procedure offers a method for using any thermistor to indicate temperature variations.

Materials

Power supply
VOM meter
Three or four spools of wire with different gauge and length
One photoresistor
One thermistor
Two resistors to be selected by student

Procedure

Part 1—The AWG Wire Table

1. Select one of the spools of wire set aside for this project and read the wire gauge number from the label. _____

2. Find both the inner and outer ends of the wire. Make sure the insulation has been stripped back about 1/4 inch from both ends. Measure the resistance of the wire from end to end with an ohmmeter. _____

3. From the wire table, find the ohms per 1000 feet for this size wire. _____

4. Calculate the length using the formula: $\text{Length in feet} = \dfrac{\text{Resistance}}{\text{Ohms}/1000'} \times 1000$

5. Now check your work! If the spool is new and not used, your calculation of length should be about the same as indicated by the manufacturer. Is it? _____

   If the spool has been used, look at what is left and "guesstimate" the amount that has been removed (1/2, 1/4, ?). _____

   Is your measurement about equal to the amount of wire remaining? _____

   There are several sources of error that could cause measurement differences. Manufacturers often add a few feet more than indicated; and, of course, if it is a partial spool, you can only estimate the actual number of feet.

6. Enter your data in Table 6-1. Complete the table for the number of spools you have available.

Table 6-1 Data for Part 1: Measurement of Wire Length

Spool A: **Calculation**
- Insulation color: _____
- Stranded or solid wire: _____
- AWG wire gauge #: _____
- Measure resistance: _____
- Length marked: _____
- Length calculated: _____

Spool B: **Calculation**
- Insulation color: _____
- Stranded or solid wire: _____
- AWG wire gauge #: _____
- Measure resistance: _____
- Length marked: _____
- Length calculated: _____

Spool C: **Calculation**
- Insulation color: _____
- Stranded or solid wire: _____
- AWG wire gauge #: _____
- Measure resistance: _____
- Length marked: _____
- Length calculated: _____

Spool D: **Calculation**
- Insulation color: _____
- Stranded or solid wire: _____
- AWG wire gauge #: _____
- Measure resistance: _____
- Length marked: _____
- Length calculated: _____

Part 2—The Photoresistor

1. Connect an ohmmeter to the leads of the photoresistor. Measure the resistance of the device in room light, in sunlight if feasible, and in darkness. (How you achieve each of these measurements is your choice.)

2. *Data*: Resistance in room light: _____
   In sunlight (if possible): _____
   With no light (dark): _____

3. When you *in*creased the light intensity, what happened to the resistance (increase, decrease, or no change)? _____

4. When you *de*creased the light intensity, what happened to resistance? _____

   A sensor is considered to have an input and an output. For this device, the input is light energy; the output is resistance.

5. From the data above, when the input increased, what happened to the output? _____

   When the input decreased, what happened to the output? _____

   An electric circuit can be used to reverse the mode of operation. As you noticed above, the output was opposite to the input. Resistance moved in a direction opposite to the change in light intensity.

   Most industrial indicators use the convention that an increasing quantity is represented by an increasing indication or measurement. Your own experience verifies this: Think of a temperature dial, fuel gauge, radio or TV sound indicator, etc.

   The photocell circuit of Figures 6-1 and 6-2 will respond to light intensity with a conventional voltmeter and will change the mode of operation to a direct indication.

   The circuit consists of the photoconductive cell in series with a 100 kΩ resistance. Voltage is measured across the resistance. As light causes the photoresistance to become smaller, the voltage across the fixed resistor will become larger. The output, therefore, follows the energy input.

   The theory and application of series connections will be studied in Experiments 8, 9, and 10.

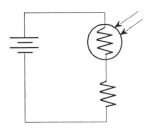
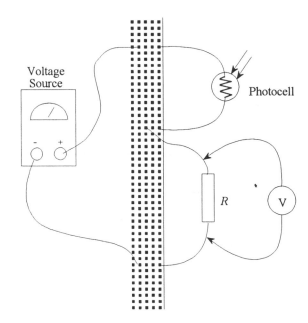

Figure 6-1   **Figure 6-2**

6. Build the light indicator circuit as shown in Figures 6-1 and 6-2.

   Source voltage is 10 V. Meter should be on 10 V range or higher.

   Change the light input to the photoresistor from maximum to dark and observe the meter indication.

   Although you are using a voltmeter, the indication represents relative light energy, and the voltage value is not important.

7. Did you achieve direct acting operation with this circuit? _____

8. You can try experimenting with different values of fixed resistance and meter scale to obtain the most effective operation. Check voltage and resistance values with your instructor.

   A manufactured light meter would have the faceplate marked in illumination units instead of volts.

9. If you placed the meter across the sensor, would the circuit action be direct or indirect?

10. Change the location of the voltmeter and confirm your answer to the previous question.

Part 3—The Thermistor Temperature Sensor

This section illustrates the operation of a semiconductor that converts temperature to resistance. The thermistor is used to measure and indicate temperature.

1. Obtain the thermistor and measure its resistance at room temperature.

2. Warm the resistor and read its resistance at elevated temperature. (How you heat the sensor is your choice, but you might ask your instructor for suggestions.)

3. What happened to the resistance as temperature increased? _____

   The thermistor is made from a semiconductor material. If an increase in temperature causes an increase in resistance, the material is said to have a *positive temperature coefficient*.

   If an increase in temperature causes a decrease in resistance, the material is said to have a *negative temperature coefficient*.

4. Based on your experimental data from steps 2 and 3, does your thermistor have a positive or negative temperature coefficient? _____

   If your thermistor creates opposite action between input and output, an electronic circuit can be used to create direct action.

5. Build a circuit similar to Part 2 to create a direct indication. Different thermistors will have different room temperature resistance. You will need to determine R for the fixed series resistance.

6. To select R, measure the thermistor resistance at room temperature. Select a series resistor approximately equal in value to the measured thermistor resistance.

7. Build a series circuit consisting of the thermistor, selected resistance, and 10 V power source.

8. Set your voltmeter on the 10 V range and place it across the series resistance.

   With this setup, ambient temperature should be indicated with the meter reading about 5 V or about center indication with an analog meter.

   The analog meter will indicate warmer temperature to the right and cooler to the left. A digital meter will indicate higher numbers for warmer temperatures and lower numbers for cooler temperatures.

   A manufactured thermometer would, of course, have a scale marked in degrees instead of volts.

9. Change the temperature input to the sensor and observe the meter indication. Does your circuit have direct or indirect action? _____

10 The Voltage Divider: Operation and Calculation

Objectives

1. Observe the voltage divider action of a series circuit.
2. Find the voltages in a series circuit without calculating current.
3. Develop skill in using the voltage divider equation.

Preparation

The voltage divider equation is an extremely useful tool in electronic technology. The principle is used in both DC and AC circuits. This project shows you how it works, saves time when calculating voltages, builds understanding of the voltage divider principle, and assists you in memorizing the equation.

Materials

DC power supply
One VOM meter
Resistors: 1/2 watt
 Four 1 kΩ, one 2.2 kΩ, one 3.3 kΩ, two 4.7 kΩ, one 5.6 kΩ, one 6.8 kΩ, one 10 kΩ, one 22 kΩ, one 47 kΩ

Procedure

Part 1—Basic Principles

1. Figure 10-1 is the circuit for the experiment. The 4.7 kΩ resistor, R_1, remains in place as you proceed and is not changed.

    The second resistor, R_2, is changed, as you progress, to each of the values indicated in the data table (Table 10-1).

    The voltage divider equation should be used for each calculation.

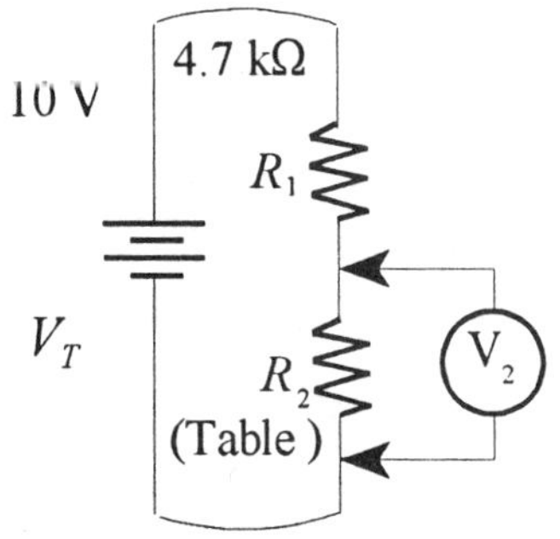

$$V_2 = \frac{R_2}{R_T} \times V_T$$

Voltage Divider Circuit

Voltage Divider Equation

Figure 10-1

2. Build the circuit with a 1 kΩ resistor for R_2.

3. Calculate the voltage across R_2 by using the voltage divider formula and enter your calculated voltage in Table 10-1.

4. Set your voltmeter to the 10 V range.

5. Hint: Use the *same voltmeter* to set the source voltage and to measure V_2. The technique of using the *measuring* instrument to adjust the source as well as measure circuit values will frequently yield more consistent results.

6. Measure the voltage across R_2. Enter in the table and calculate error.

7. Change R_2 to the next higher resistance value in the chart.

8. Since you have increased the resistance of R_2, do you expect the voltage across R_2 to go up, down, or stay the same? _____

9. Measure the voltage across R_2. _____

10. What happened to V_2? _____

11. Complete the calculations and measurements for the remainder of the table. Remember to use the voltage divider equation. It will become easier to use as you progress.

    Are your errors satisfactory? _____

Table 10-1 Data for Figure 10-1

R_2	Calculate Total Circuit Resistance	Calculate V_2	Measure V_2	% Lab Error
1 kΩ				
2.2 kΩ				
3.3 kΩ				
4.7 kΩ				
5.6 kΩ				
6.8 kΩ				
10 kΩ				
22 kΩ				
47 kΩ				

Part 2—Equal Resistors and Ratio Method

The same principles apply to any circuit whether two, three, or more resistances are in the series string.

If all the resistances are equal, the total voltage will divide equally across each resistor. In a string of three equal resistors, each component will drop 1/3 of the source voltage. In a 4-resistor circuit, each resistor will drop 1/4 of the source. Two resistors will drop 2/4 or 1/2.

A simple ratio or fraction, therefore, will calculate the voltage across one or more components.

1. Build the circuit of Figure 10-2. All resistors have the same value.

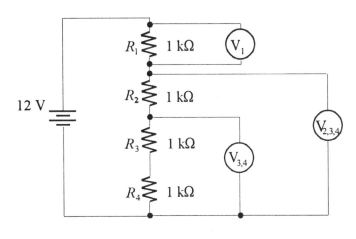

Figure 10-2

2. Measure the voltage across R_3 and R_4 combined.

   Since you are measuring the voltage across two out of four equal resistances in series, the voltage should be equal to 2/4 or 1/2 of the source voltage.

   Do you measure 1/2 of 12 V, or 6 V, across R_3 and R_4? _____

3. What fraction of the total voltage should you measure across a single resistance such as R_1?

   _____

   Measure V_1 and compare with your calculation. _____

4. How much voltage should you measure across three resistors? _____

5. Measure the voltage across the combination of R_2, R_3, and R_4. _____

6. Enter your calculations and measurements in Table 10-2. Compare results for verification.

Table 10-2 Data for Figure 10-2

Part of Circuit Measured	Calculate Voltage Across	Measure Voltage Across	Compare Calculation and Measurement
R_1 only			
R_3 and R_4			
R_2, R_3, and R_4			

Part 3—Unequal Resistances in Series: Voltage Division

Many circuits have resistance and voltage values that are not equal. Refer to Figure 10-3. Here you can use the voltage divider equation from Part 1.

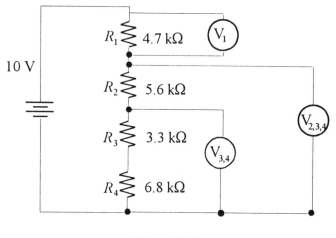

Figure 10-3

1. Calculate the voltage across R_1 using the voltage divider equation. _____

2. Build the circuit and measure the voltage across R_1. _____

3. Calculate the voltage across the combination of R_3 and R_4. You will need to interpret the equation as follows:

$$V \text{ across any part of the circuit} = \frac{\text{Resistance of that part of the circuit}}{\text{Resistance of the whole circuit}} \times V_T$$

$V_{3,4} =$ _____

4. Measure the voltage across the combination of R_3 and R_4 and compare with your calculation. _____

5. What voltage divider equation could you use to calculate the voltage across the combination of R_2, R_3, and R_4? Write your equation here, based on the principles from steps 1 and 3.

6. Combine your data below. Compare results and complete the table.

Table 10-3 Data for Figure 10-3

Part of Circuit Measured	Calculate Voltage Across	Measure Voltage Across	Compare Calculation and Measurement
R_1 only			
R_3 and R_4			
R_2, R_3, and R_4			

Lab Questions

Sometimes an experiment can be done with paper and pencil as well as hardware. Here is a "paper experiment." Use the ratio, or voltage divider method, to answer questions.

1. Refer to the circuit of Figure 10-4. What *fraction* of the total voltage will appear across R_2? _____

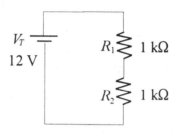

Figure 10-4

2. Replace R_2 with a 2 kΩ resistor. Draw the new circuit here. Include resistance values and source voltage.

3. What voltage will you measure across R_2 for the circuit of question 2? _____

4. Put back the original 1 kΩ resistor at R_2 and replace R_1 with a 0.5 kΩ resistor.

    You will now have: $R_1 = 0.5$ kΩ; $R_2 = 1$ kΩ

    Draw this circuit here. Include resistance values and source voltage.

5. What voltage will you measure across R_2 for the circuit of question 4? _____

6. Why do you have the same voltage across R_2 for each of the last two circuits?

11 Parallel Circuits 1: Normal Operation

Objectives

1. Observe the operation of a parallel circuit.
2. Demonstrate a method for measuring total circuit resistance for any circuit.
3. Observe the effect of adding an additional parallel component on circuit currents and resistance.

Preparation

The experiment looks at the operation of parallel circuits and also presents several laboratory measurement techniques that may be useful here and in future projects.

Materials

DC power supply
VOM meter
Resistors: 1/2 watt
   One 1 kΩ, two 2.2 kΩ

Procedure

Part 1—Circuit Resistance

1. Build the circuit of Figure 11-1.

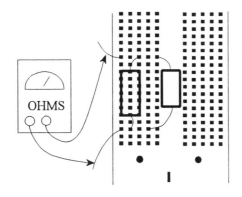 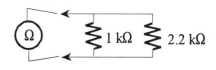

Figure 11-1

2. Calculate the total circuit resistance. _____

3. Measure the total circuit resistance. _____

   (Here is an effective technique for measuring R_T for any circuit: Disconnect the power supply leads at the power source. Connect the ohmmeter to the ends of leads that were connected to the voltage source. This method will always read R_T for the circuit.)

4. Enter your results in Table 11-1 and calculate the laboratory error.

5. If you connect another resistance in parallel with the first two, what will happen to R_T (increase, decrease, or no change)? _____

6. Add another 2.2 kΩ resistance in parallel with the first two and measure the new value of R_T.

   _____

7. What happened to the circuit resistance when another resistor was *added* in parallel?

   _____

   A note on experimental technique for step 7: Connect and disconnect one lead of the third resistor while looking at the meter. You can directly observe the effect of this component on circuit measurements.

Part 2—Measurement of Branch and Line Currents

1. Wire the parallel circuit shown in Figure 11-2. The jumper wires are used to assist in connecting the ammeter.

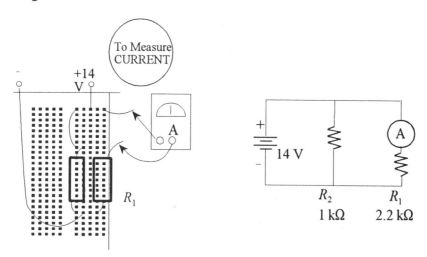

Figure 11-2

2. Calculate I_1, I_2, and I_T. Enter these values in Table 11-1.

3. Measure current through R_1 (Figure 11-2).

   Method: To measure I_1, lift the bottom lead of the jumper wire and the top lead of R_1. Connect the ammeter to these two leads and read the current value.

4. Measure I_2 the same way and enter both readings in the data table.

5. Measure total circuit current, I_T. (This is also called line current.)

   Method: To measure I_T, disconnect the positive lead *at* the voltage source. Connect the ammeter between the positive terminal of the voltage source and the wire lead you just disconnected.

   If the ammeter is in series with the power supply and the circuit board, it will be measuring circuit current.

   Enter I_T in the data table and calculate percent error.

6. If you add another resistor in parallel, what will happen to the line current? _____

7. Place another 2.2 kΩ resistor in parallel with the first two. Observe the line current.

   _____

   Once the resistor is in place, use the technique of making and breaking the connection of the third resistor while watching the ammeter. Check the meter scale to be sure it does not overrange.

8. Does I_T increase or decrease with the addition of another resistance in parallel to the circuit?

   _____

   For the following steps, be sure the source voltage does not change. A regulated supply would be ideal. Otherwise, check voltage with a meter and adjust if necessary.

9. What happens to the current through the 1 kΩ resistor when you add a third resistor in parallel (increase, decrease, or no change)? _____

10. Set up the circuit with the ammeter measuring I_1 as in Figure 11-2. Add a 2.2 kΩ resistor in parallel with the first two resistors. Observe the effect on I_2 by making and breaking the connection of the added resistance.

    If the circuit is connected properly, there should be no change in current through the original resistors when adding another in parallel.

Part 3—Voltage Across Each Parallel Branch

1. Disconnect the ammeter and replace the jumper wires. (Put the circuit back in its original condition with no meters attached.)

2. Adjust the meter to measure voltage (Figure 11-3).

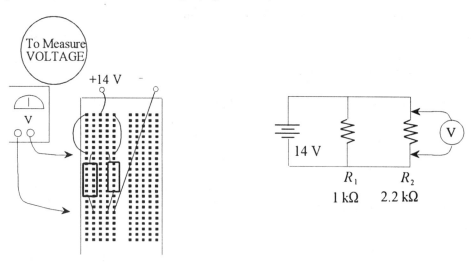

Figure 11-3

3. Measure the voltage across R_1. _____

   Measure the voltage across R_2. _____

4. Does the same voltage exist across both resistors in a parallel arrangement? _____

5. If you add a third resistor in parallel, will this additional resistor affect the voltage across the first two? _____

   Will it have the same voltage as the first two? _____

6. Add another 2.2 kΩ resistor in parallel with the first two.

7. Measure the voltage across each resistor.

   Are all voltages exactly the same? _____

Table 11-1 Data for Parallel Circuits: Figures 11-1, 11-2, 11-3

Circuit	Unit	Calculate Value	Measure Value	% Difference
11-1	R_T			
11-2	I_1			
11-2	I_2			
11-2	I_T			
11-3	V_1			
11-3	V_2			
11-3	V_3			

Lab Questions

1. Why did the total parallel resistance become smaller when you added a third resistance in parallel?

2. What would have happened to the circuit resistance if you removed a resistor instead of adding?

3. In Part 2, why did total current become larger when you added a third resistance in parallel?

4. What will happen to circuit current if you remove a parallel resistance? Explain the reason this happens.

5. Why did the current through I_1 not increase when you added another parallel R?

12 Parallel Circuits 2: The Effect of Opens and Shorts

Objectives

1. Observe the effect of opens and shorts on a parallel circuit.
2. Predict the result of these problems on circuit operation.

Preparation

In this project, you will set up a properly working parallel circuit and measure the values of current and voltage for reference. Next, you will create problems, such as open circuit or shorted component, and measure the effect of these problems on circuit operation.

The values you will determine in Part 1 are the reference quantities. They will be used as a bench mark to compare with the results of problems created in Parts 2-6.

Materials

DC power supply
VOM meter
Resistors: 1/2 watt
   One 1 kΩ, one 2.2 kΩ, one 10 kΩ

Procedure

Part 1—Normal Circuit Values for Reference and Comparison

1. Wire the parallel circuit of Figure 12-1. This is the same circuit used in the previous project. Use the same technique to measure currents.

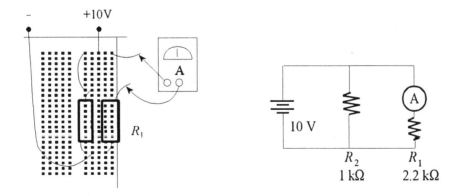

Figure 12-1

2. Measure: $I_1 =$ _____ $I_2 =$ _____ $I_T =$ _____

3. Does the sum of the two branch currents add up to the value of I_T? _____

   If they do, it indicates your measurements are probably correct. If they do *not*, then you need to check your project. Check resistor values, meter placement, or circuit layout.

4. Measure: V_1 _____ V_2 _____ V_T _____

   Are these *exactly* the same? _____ Should they be *exactly* the same? _____

5. Disconnect circuit from power source and measure total resistance. _____

Part 2—The Effect of a Short Circuit on Total Circuit Resistance

1. What do you expect will happen to the total circuit resistance if a shorting wire is placed across one of the branches (increase, decrease, no change, become zero, become infinite)?
   _____

2. Take a small jumper wire, stripped on both ends, and connect it directly across the leads of the first resistor. Be certain the circuit is disconnected from the power source and measure the new R_T. _____

3. Place the shorting wire directly across the second resistor and measure the circuit R_T.
   _____

4. Plug the shorting wire into the board in parallel with the combination and measure the R_T.
   _____

   Analysis: Based on the results of your experiment above, a short circuit across *any* component in a parallel combination makes the resistance of the entire combination equal to _____ ohms.

Part 3—The Effect of an Open Branch on Total Circuit Resistance

1. What will happen to the circuit resistance if one parallel component is opened (increase, decrease, become zero, become infinite)? _____

2. Connect the ohmmeter to the input leads and take a reading of R_T. _____

   (Remember to disconnect the input leads from the voltage source.)

3. With the ohmmeter still connected to measure R_T, *open* the first resistor by lifting one lead and read the new R_T. _____

   What happens to circuit resistance when one branch opens? _____

4. Replace the lead of the first resistor and open the second resistor.

   Measure the new R_T. _____ How does this compare with the original resistance? _____

Part 4—The Effect of an Open Branch on Total Circuit Current

1. Wire the ammeter in series with the source. (This measures I_T.)

2. Apply 10 V and read the total current. _____

3. Open one resistor by lifting a lead off the board and read the new circuit current.

   What happens to total current when one branch opens? _____

Part 5—The Effect of an Open Branch on Other Branch Currents

1. Wire your ammeter in series with the first resistor.

2. What do you think will happen to the current through this first resistor if the other resistor should open? _____

3. Open the second resistor and observe the current flowing through the first. _____

   What happened to the current through the first resistor when you *opened* the other branch? _____

Part 6—The Effect of Adding Another Parallel Resistance on Branch Currents

1. Leave your ammeter connected as in Part 5 (measuring current through R_1). Be sure the two branches are connected in parallel.

2. Place a third resistor (10 kΩ) in parallel with the first two. Connect and disconnect this resistor and observe the effect on the current flow through the R_1.

   Does I_1 change with the addition of the third resistor? _____

   Note: With a well-regulated voltage source, there should be no observable change in current flow through I_1. If the source is not regulated, there may be some small change due to the loading of the power supply.

Lab Questions

1. In Part 2, you placed a short across one of the resistors. Was this shorting wire in parallel with the circuit, in series, or in no particular arrangement?

2. According to the principles of parallel resistance, the total circuit resistance is less than the ohms of the smallest branch.

   Use this principle to explain why a short across one branch caused R_T to become zero.

3. In Part 3, you opened a resistor by lifting one lead from the board, leaving the other end connected.

   In the previous experiment, the same result occurred by removing the resistor completely from the board. Why does removing a resistor completely create the same result as opening one lead?

4. Opening one resistor did not affect the current through the other branches. Why?

5. Adding a series resistance to a series circuit caused the circuit current to *decrease*.

   Adding another parallel resistance to a parallel combination caused the circuit current to *increase*. Why?

13 Series-Parallel Circuits 1: Normal Operation

Objectives

1. Build a series-parallel circuit on the breadboard.
2. Obtain current and voltage measurements.
3. Observe Kirchoff's voltage and current laws applied to a series-parallel circuit.

Preparation

The first circuit is a simple 3-resistor circuit. Although there are only three components, the elements of operation are the same as occur in a larger circuit. You will begin by making theoretical calculations for each element, and then measure the electrical quantities.

The second circuit contains more components and is used to demonstrate Kirchoff's voltage and current laws. It also creates additional practice in circuit measurements and calculations.

Materials

DC power supply
One VOM meter
Resistors: 1/2 watt
   One 1 kΩ, one 2.2 kΩ, two 4.7 kΩ, one 5.6 kΩ, one 6.8 kΩ

Procedure

Part 1—Circuit 1

1. The first circuit is Figure 13-1. Look at the data sheet, Table 13-1. The table can be used as a guide for performing the experiment. The calculation column should be done first in this project. Calculate each of the values indicated and enter your answers in column 2.

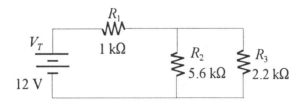

Figure 13-1 Circuit 1

2. Build the circuit. Component layout should be similar to the schematic. It is not necessary that the physical geometry match the diagram exactly, but it can help in performing measurements and circuit analysis.

3. Next, measure each voltage and current and enter these measurements in the table.

4. To find the power generated by each component, multiply the *measured* voltage times the *measured* current. This will be the *measured* power. Enter these values in the third column. Units should be milliwatts (mW).

   Hint: Multiplying volts times milliamps yields milliwatts for the unit. Example: 4 V × 2 mA would equal 8 mW for the third column.

5. Compare calculated and measured data by calculating a percent difference. Use the equation from Experiment 5.

Table 13-1 Data for Circuit 1, Figure 13-1

1	2	3	4
Quantity	Calculation	Measurement	Compare Columns 2 and 3
V_T			
R_T			
I_T			
V_1			
V_2			
V_3			
I_1			
I_2			
I_3			
P_1			
P_2			
P_3			
$P_T = P_1 + P_2 + P_3$			
$P_T = V_T \times I_T$			

Part 2—Circuit 2

1. Start by calculating the electrical values for Circuit 2 as indicated in Table 13-2.

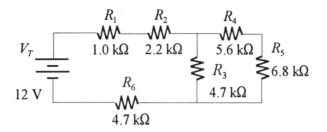

Figure 13-2 Circuit 2

Enter your calculations in the second column.

2. Build Circuit 2 and measure the values indicated. Enter your measurements in the third column.

3. Compare your calculated and measured data by finding the percent difference.

Table 13-2 Data for Circuit 2, Figure 13-2

1 Quantity	2 Calculation	3 Measurement	4 Compare Columns 2 and 3
V_T			
R_T			
I_T			
V_1			
V_2			
V_3			
V_4			
V_5			
V_6			
I_1			
I_2			
I_3			
I_4			
I_5			
I_6			

Lab Questions

Circuit 1

Before answering the following questions, make sure that your calculations and measurement data are sufficiently close and within experimental tolerance. If they are not close enough, you need to check your calculations, measurements, circuit wiring, or components.

1. Can you find the total power developed in a circuit by adding the individual powers for each component? _____

   Does $P_T = P_1 + P_2 + P_3$? _____

2. Verify your answer to question 1 from the measured data. Find the circuit power by the relationship $P_T = V_T \times I_T$. _____

   Add the measured values $P_1 + P_2 + P_3$. _____

   Are your two calculations the same—within lab tolerance? _____

3. Look at the schematic for Circuit 1. According to Kirchoff's current law, should $I_2 + I_3 = I_1$?
   _____

4. Find out if your data agree with your answer to 3. Add your *measured* values of $I_2 + I_3$.
   _____

5. Why are V_2 and V_3 exactly the same voltage? _____

6. One side of the 1 kΩ resistor is connected directly to the positive side of the 12 V power source. Why doesn't the voltage *across* this resistor measure 12 V?

7. Can you think of a way to determine the current through a resistor without using an ammeter? Describe the method.

Circuit 2

According to *Kirchoff's Voltage Law*, the sum of the voltage drops around any loop in a circuit will equal the source voltage.

Definition: A "loop" is any path that current can take, moving from one side of the voltage source to the other. There are usually several paths in a circuit, and they can overlap in some locations.

1. Look at Circuit 2. One loop would consist of the current path through R_1-R_2-R_3-R_6. Trace this loop in pencil on the circuit diagram.

2. Add the voltage drops across each resistor in this loop. Does $V_1 + V_2 + V_3 + V_6 = V_T$? _____

3. Name the resistors for another loop in Circuit 2. _____

4. Trace this second loop in pencil on the circuit diagram.

5. Add the voltages in the loop consisting of V_1, V_2, V_4, V_5, and V_6. _____

   Is the sum of the voltages in this loop equal to source voltage (within normal experimental tolerance)? _____

6. Do your data illustrate Kirchoff's voltage law? _____

7. According to *Kirchoff's Current Law*, the sum of the currents flowing into a point will equal the currents flowing away from that point. Do your data illustrate this rule? _____

   Does $I_6 = I_3 + I_5$? _____

   Does $I_2 = I_3 + I_4$? _____

8. Why is the current flowing through R_4 and R_5 the same?

9. R_6 and R_3 are *both* 4.7 kΩ. Why do they have different voltage drops?

10. From your data, does $V_4 + V_5 = V_3$? _____ Why?

11. Why do I_1, I_2, and I_6 all have the same current?

14 Series Parallel Circuits 2: The Effect of Opens and Shorts

Objectives

1. Build a complex series parallel circuit. Analyze the operation of the circuit through calculation and measurement.
2. Build an equivalent circuit (sometimes called simplified or reduced) to assist in analyzing and troubleshooting the complex circuit.
3. Create opens and shorts in a series parallel circuit and observe the effect of these problems on the circuit operation.

Preparation

In Part 1, you will build a complex series parallel circuit and observe its operation through calculation and measurement. These values will be used as a base, or reference point, for further activities.

In Part 2, you will calculate and build a simplified or equivalent version of the circuit. This will be used to illustrate the advantages of the "equivalent" concept for analysis and troubleshooting.

In Parts 3 and 4, you will create opens and shorts and observe the effect of these problems on the circuit.

Materials

One DC voltage source
One VOM meter
Resistors: 1/2 watt
    Three 10 kΩ, one 4.7 kΩ, one 3.3 kΩ

Procedure

Part 1—Test Circuit

1. Figure 14–1 is the series/parallel circuit for the experiment.

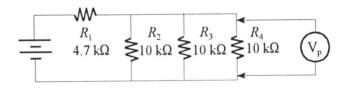

Figure 14–1

2. Calculate R_T, I_1, V_1, and V_P. Enter the results in Table 14–1. (V_P is the voltage across the three parallel branches.)

3. Build the circuit. Measure the values indicated and enter your data in Table 14–1.

Table 14–1 Data for Original Test Circuit, Figure 14–1

Quantity	Calculate	Measure	Compare
R_T			
I_1			
V_1			
V_P			

Part 2—Equivalent Circuit

1. Note the circuit of Figure 14–2. It replaces the three parallel resistors with a single equivalent resistor. This single component has the same resistance value as the parallel combination.

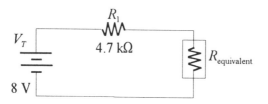

Figure 14–2

2. Calculate the combined value of the three 10 kΩ parallel resistors. _____

3. Remove the three 10 kΩ resistors from the original circuit and replace them with a single equivalent resistor whose value you calculated in step 2.

4. Apply power to the circuit and measure the voltage across R_{eq}. _____

   Does V_{eq} have the same voltage as V_P in the original circuit (within experimental tolerance)? _____

   (It should. If your results are very different, check wiring, color code, or calculation.)

5. Measure the total circuit resistance. (Remember, R_T is the resistance between the positive and negative supply leads *with power source disconnected.*) _____

6. Is R_T for the equivalent circuit the same resistance as R_T for the original circuit? _____

7. Complete the data for Table 14–2. Take the measurements and make the calculations for the equivalent circuit. Compare your results.

Table 14-2 Data for Equivalent Circuit, Figure 14-2

Quantity	Calculate	Measure	Compare
R_T			
I_1			
V_1			
V_P			

8. Transfer data to Table 14-3 to compare measurements between the original test circuit and the simplified equivalent circuit.

Table 14-3 Comparison of Original and Equivalent Circuits

Quantity	Original Circuit Measurement	Equivalent Circuit Measurement	Comparison
R_T			
I_1			
V_1			
V_P			

Part 3—Open Circuit Problems

1. Rebuild the original circuit (Figure 14-1). Do not connect the power leads at this time.

2. What will happen to the total circuit resistance if R_3 should open? (Will R_T increase, decrease or not change?) _____

3. If R_3 should open, what would happen to the equivalent resistance of the parallel branches? _____

4. Lift one lead of R_3 from the board, simulating a break or "open" in R_3.

5. Measure the new resistance for R_T with R_3 open. _____

6. Since the power leads are not connected at this time, an ohmmeter connected across R_4 will show the new equivalent resistance of the three parallel branches.

   Measure the equivalent resistance of R_P with R_3 open. _____

7. Were your predictions correct? _____

8. Close R_3 by connecting it to the board and apply the 12 V power source.

9. With power on, measure: V_1 _____, I_1 _____, and V_P _____.
    (These measurements represent the values of the properly working circuit.)

10. Lift one lead of R_2 and observe the voltage across R_1. What happened to V_1? _____

11. Based on the results of V_1, what happened to I_1 when you opened R_2? _____ What happened to circuit current? _____

12. Measure the voltage across R_3. Is it more or less voltage than you measured in the good circuit? _____ What caused this change in voltage?

13. If *all three* parallel resistors open, what will happen to the voltage across V_P?

14. If *all three* parallel resistors open, what will happen to the voltage across V_1?

15. Open all three parallel resistors and measure V_P. _____ Measure V_1. _____

16. Connect the parallel resistors and open R_1.

17. Predict the voltage across V_3. _____ Measure V_3. _____

18. Predict the voltage across V_1. _____ Measure V_1. _____

Summarize the results of your "open circuit" problems by completing Tables 14-4 and 14-5. Make any measurements or calculations you may need to complete the data.

Table 14-4 Open Circuit Problem: Open R_2

Quantity	Good Circuit Measurement	Predicted Value with Open R_2	Measured Value with Open R_2
R_T			
I_1			
V_1			
V_P			

Table 14-5 Open Circuit Problem: Open R_1

Quantity	Good Circuit Measurement	Predicted Value with Open R_1	Measured Value with Open R_1
R_T			
I_1			
V_1			
V_P			

Part 4—Short Circuit Problems

This section looks at the effects of shorted circuit components. You can create this effect by placing a piece of hook-up wire in contact with the leads of a given component.

    Tables 14-6 and 14-7 can be used as a guide for this section. The quantities to be checked are indicated in the first column.

    Make your prediction for the results of the problem and enter in the third column.

    Short out R_4 or R_1 and measure the results. Complete the tables as indicated. Were your predictions correct? _____

Table 14-6 Problem: Shorted R_4

Quantity	Good Circuit Measurement	Predicted Values with Shorted R_4	Measured Values with Shorted R_4
R_T			
R_P			
V_1			
I_1			
V_P			

Table 14-7 Problem: Shorted R_1

Quantity	Good Circuit Measurement	Predicted Values with Shorted R_1	Measured Values with Shorted R_1
R_T			
R_P			
V_1			
I_1			
V_P			

Lab Questions

1. What is the advantage in using "equivalent" arrangements to analyze circuits?

2. What is the circuit form (series or parallel or series/parallel) for the equivalent circuit in this experiment?

3. Why did R_P increase when R_3 opened?

4. Why did R_T increase when R_3 opened?

5. Using your answers to questions 3 and 4, explain why circuit current became smaller when R_2 opened.

6. What voltage appeared across R_P when all three parallel resistors were open? Why?

7. Did you notice that the measured V_P was actually somewhat *less* than source voltage in question 6? If so, what characteristic of the voltmeter, as part of the circuit, caused this to happen?

8. Why did zero voltage appear across R_1 when all three parallel resistors were open?

9. Why did source voltage (V_T) appear across R_P when R_1 was shorted?

10. Why did source voltage appear across R_1 when R_4 was shorted?

15 Voltmeter and Circuit Loading

Objectives

1. Observe the effect of voltmeter loading on circuit measurements.
2. Calculate measurement indication by combining the meter's input resistance with component values.
3. Analyze experimental data to determine the effect of meter sensitivity and scale selection on voltage measurement.

Preparation

Have you noticed in past experiments that some voltage measurements are lower than expected even though you checked everything very carefully?

Have you noticed on occasion that you get a certain voltage reading on one range, but when you change scales, the reading is *different*?

This lab demonstrates the reason these situations occur and suggests methods for avoiding them. It is designed primarily for analog meters with DC input impedance of 20 kΩ/V. The last section does allow you to use a digital meter or analog meter with greater sensitivity to compare results.

The principles learned here will apply also to AC measurements and other instruments such as the oscilloscope and AC meters.

Technicians often use measuring instruments as if they were perfect devices that do not affect the circuit. In most cases, this is a reasonable assumption.

But meters do have resistance, and as soon as they are connected to the circuit, change the circuit because of this resistance.

Example: Refer to Circuit A (Figure 15-1) below.

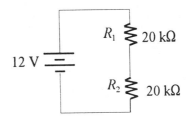

Figure 15-1 Circuit A

Both resistors are the same value. The theoretical and actual voltage across R_2 is 1/2 the total, or 6 V.

Refer to the next diagram, Circuit B (Figure 15-2). There is a third resistance in parallel with R_2, which, of course, changes the circuit.

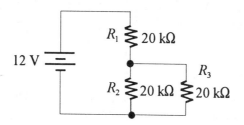

Figure 15-2 Circuit B

The equivalent Circuit C (Figure 15-3) combines the two parallel resistances into a single R_P, which is 10 kΩ.

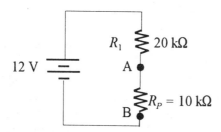

Figure 15-3 Circuit C

If you calculate the voltage between A-B for Circuit C, it will have 1/3 of the total voltage, or 4 V. This is less than the original circuit. The addition of the third resistance changed the circuit and lowered the voltage across R_2.

The same situation happens when a voltmeter is attached to a circuit. Refer to the original Circuit A, but this time assume there is a 20 kΩ/V meter across R_2. If the range selector is on the 10 V range, the meter's input resistance will be (20 kΩ × 10) or 200 kΩ. So you are really placing a 200 kΩ resistance in parallel with the second resistance. This creates Circuit D (Figure 15-4).

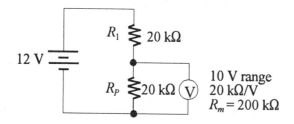

Figure 15-4 Circuit D

The parallel value of 20 kΩ and 200 kΩ = 18.18 kΩ.

The voltage across the parallel combination is 5.714 V, not 6 V. This situation is called *voltmeter loading*. It is one more item that contributes to overall laboratory error.

At this point, you might want to be sure that you can do the calculations shown above. You will need to do them as the experiment progresses.

Do your own calculations for the examples above and compare results.

Materials

Regulated voltage source
Analog voltmeter: 20 kΩ/V
Resistors: 1/2 watt
  Two 5.6 kΩ, two 56 kΩ, two 560 kΩ, two 1 MΩ
  Optional: Two 10 kΩ, two 100 kΩ, two 2.2 MΩ

Procedure

Part 1—Meter Data

1. Start by listing the sensitivity, or ohms/volt rating, for your particular meter. 20 kΩ/V is very common, but your meter might have a different value.

   Meter name: _____ Model # _____ Ohms/volt: _____

2. Calculate your meter input resistance (in kilohms) for the 2.5 V range (or similar range).

3. Calculate your meter input resistance (in kilohms) for the 10 V range.

4. Enter both of these values at the bottom of Table 15-2.

5. You can *measure* the meter's input resistance as well as calculate the value. You will need to use two meters: one set as an ohmmeter, the other as a voltmeter.

   Select the 10 V scale on the voltmeter and measure the resistance at its input terminals.

   _____

   Is this the same input resistance you calculated in step 3? _____

6. You can measure meter resistance on the 2.5 V scale as well. However, this connection may overrange the voltmeter because of the power supplied by the ohmmeter. Check with your instructor on the feasibility of this particular measurement.

Part 2—Circuit Loading

1. Refer to Figure 15-5. The circuit consists of a simple two-resistor series connection. R_1 and R_2 are the same value. After taking data for one pair of resistors, you will change their value to the next set of resistors.

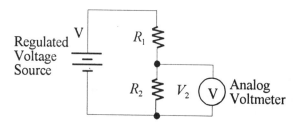

Figure 15-5

2. Build the circuit (Figure 15-5). R_1 and R_2 are 5.6 kΩ each.

3. Table 15-1 is data for 4 V applied. Calculate the theoretical voltage across R_2. Since both resistances are the same, V_2 is 1/2 the source voltage. Enter this in column 2.

4. Measure the voltage across R_2 with your voltmeter on the 2.5 V range. Enter this value in column 3.

   A suggestion on technique: Use the measuring voltmeter to *set* the source voltage. Laboratory meters are often more accurate and repeatable than power supply or "built-in" meters.

5. Calculate the percent error between theoretical and measured values. Enter this difference in column 4.

   You will notice some very large differences in this column! You may not believe the results at first, but keep in mind that this particular experiment is designed to exaggerate meter effects on the circuit.

   Note: The meter is reading the *true* voltage. It is showing you what is across the resistance at the time of measurement. Since the circuit has been changed with the addition of the meter, the actual voltage has changed also. Meter accuracy is not a major factor here, although it may contribute a small amount to the indication.

6. Calculate the true voltage appearing across R_2 by including the meter resistance in the calculation. This is done using the same method illustrated in the Preparation section.

   When you include the meter resistance in parallel with R_2 as part of your calculation, you will notice that measured and calculated values are much closer. They should, in fact, be the same.

7. Increase the source voltage to 16 V and complete Table 15-2. Place the range selector on the 10 V range and then increase the supply voltage to 16 V.

8. The third table (Table 15-3) is optional and is included to obtain data for an electronic or digital voltmeter. These have very high input resistances, typically on the order of 11 MΩ.

Table 15-1 Data for Source Voltage = 4 V, 20 kΩ/V Meter Set on 2.5 V Range

1	2	3	4	5
R_1 and R_2 Circuit Values	Calculate Theoretical V_2	Measure V_2 with Meter	Calculate the % Error Between Columns 2 and 3	Calculate V_2 Including Meter Ohms
5.6 kΩ				
56 kΩ				
560 kΩ				
1 MΩ				

Table 15-2 Data for Source Voltage = 16 V, 20 kΩ/V Meter Set on 10 V Range

1	2	3	4	5
R_1 and R_2 Circuit Values	Calculate Theoretical V_2	Measure V_2 with Meter	Calculate the % Error Between Columns 2 and 3	Calculate V_2 Including Meter Ohms
5.6 kΩ				
56 kΩ				
560 kΩ				
1 MΩ				

Meter Name: _____
Sensitivity in Ohms/Volt: _____

Table 15-3 Data for Source Voltage = 16 V, Electronic Digital or Analog Meter

1	2	3	4	5
R_1 and R_2 Circuit Values	Calculate Theoretical V_2	Measure V_2 with Meter	Calculate the % Error Between Columns 2 and 3	Calculate V_2 Including Meter Ohms
10 kΩ				
100 kΩ				
1.0 MΩ				
2.2 MΩ				

Meter Name: _____   Input Impedance: _____

Lab Questions

1. Refer to column 3, Table 15-1. When did the meter affect the circuit most—when reading voltage across higher resistance or lower resistance? _____

2. Refer to Table 15-2, column 3. The range selector was set to a higher scale. Did this make the input resistance larger or smaller? _____

3. How did the meter range setting affect the resistance between points A-B?

4. What caused the voltmeter to have a smaller *loading* effect in Table 15-2?

5. Does a meter with a higher "sensitivity" have more or less input resistance? _____

6. A student is measuring voltage with a 20 kΩ/V meter on the 10 V range. She switches the meter selector to the 2.5 V range and notices a drop in voltage reading.

    What is the cause of this change in voltage reading?

7. If you used a meter with a higher ohms/volt rating, would you expect "better" or "worse" readings? _____

8. Refer to your calculations in column 5 of either Table 15-1 or Table 15-2.

    When you included meter resistance in your calculations, did you get results closer to the measured value as compared to the theoretical values in column 2? _____

9. (Optional) If you used a different meter for Table 15-3, list the:

    Meter Name: _____

    Type (Digital, Analog with FET transistor input, etc.): _____

    DC input resistance: _____

10. Is the input resistance of the electronic meter constant for all ranges, or does it change with range setting? _____

16 The Variable Resistor

Objectives

1. Demonstrate the operation of the potentiometer as a variable resistor.
2. Observe the effect of terminal connections on resistance change.

Preparation

As with all projects, you need to have studied the device prior to performing the experiment. You will operate the variable resistor and observe the resistance changes at the terminals.

Since there are many forms and sizes of variable resistors, the components indicated here are only representative. The characteristics of resistance change are common, but connection pins and locations may vary.

A PC board connection is shown in Figure 16-1. The mounting position is important.

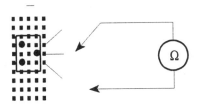

Figure 16-1

The device has the generic name "potentiometer."

The terms *variable resistor, potentiometer*, or *rheostat* are used, depending on the circuit configuration.

Since its purpose is to "control" an electronic function, such as volume, brightness, motor speed, etc., the generic term "control" is sometimes used as well.

Materials

One VOM meter
One PC potentiometer

Procedure

1. Select a 10 kΩ PC variable resistor. Any other reasonable value will work as well. Notice the pin numbering. Most are marked with pin 1 and pin 3 indicated. These are the outer terminals. The center is the variable wiper, or pin 2. The ohms value is usually stamped or printed on the edge.

2. Mount the variable resistor on your breadboard as shown in Figure 16-1.

3. Connect an ohmmeter to terminals 1 and 3 (the outside terminals).

4. Measure the total resistance of the device.

    Is this the same value as marked or stamped on the part? _____ Resistor tolerance values for potentiometers are often as much as 20%.

    Turn the wiper. Does the ohmmeter reading between terminals 1 and 3 change? _____
    Why not?

5. Change the ohmmeter connections to terminals 1 and 2.

6. Turn the wiper slowly *clockwise*. Does the resistance *increase* from zero to full resistance? _____

    Turn the wiper slowly *counterclockwise*. Does the resistance *decrease* from maximum ohms to zero? _____

7. Change the ohmmeter connection to terminals 2 and 3.

8. Turn the wiper clockwise. What happens to the resistance? _____

9. Turn the wiper counterclockwise. What happens to the resistance? _____

10. How does the choice of terminal connections affect the resistance change?

17 The Potentiometer and Voltage Control

Objectives

1. Demonstrate the operation of the potentiometer as a voltage control.
2. Connect the voltage source and load resistance for proper operation.

Preparation

Mount the potentiometer as shown in Figure 17-1. The connecting wires let you make contact with the terminals without raising the potentiometer (pot.) from the board.

Note: Be certain the power source is connected to terminals 1 and 3 and not inadvertently connected to the center terminal.

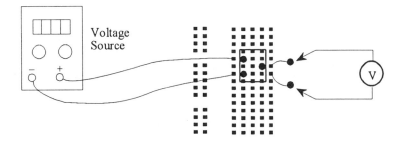

Figure 17-1 The Potentiometer as a Voltage Control

Materials

Power supply
Voltmeter
One 10 kΩ PC potentiometer

Procedure

1. Set the power supply to 10 V. Select the 10 V range on the voltmeter. Terminals 1 and 3 are considered the *input* to the device. Terminals 1 and 2 are considered the *output* from the device.

2. Check the circuit operation. Slowly turn the wiper clockwise. Does the voltage output change as you adjust the wiper? _____

   If it does not change, check your voltmeter connection. Be sure it is on terminals 1 and 2 and not 1 and 3.

3. Turn the adjustment fully clockwise. Measure the output voltage. _____

4. Turn the adjustment fully counterclockwise. Measure the output. _____

5. According to your data, the output voltage on terminals 1 and 2 will (increase or decrease) _____ as the adjustment is turned *clockwise*.

6. Disconnect both voltmeter leads and reconnect them to terminals 2 and 3 (positive on 3 and negative on 2).

7. Once again turn the adjustment clockwise. Does the voltage increase or decrease with *clockwise* adjustment? _____

8. From your data, the output voltage at pins 2 and 3 will _____ as the adjustment is turned clockwise.

9. One application for the potentiometer is the volume control for radio or TV. As you know, the volume increases with clockwise rotation or adjustment.

   For this operation, which terminals connect to the input? _____

   Which terminals connect to the output for proper operation? _____

18 The Rheostat and Current Control

Objectives

1. Build a rheostat circuit to control current.
2. Calculate and demonstrate the maximum and minimum controllable current.

Preparation

The circuit uses a 2.2 kΩ resistor for load, a 10 kΩ 1/2 W potentiometer, and an ammeter to indicate current flow.

Calculation: You will need to calculate the maximum and minimum currents that the circuit can control for the load involved. Here is how it is done:

Maximum current: Refer to Figure 18-1. If the wiper is moved to its extreme left position, it presents zero resistance to the series circuit. This creates the least total resistance, so there will be maximum current. The current is calculated by dividing the source voltage by the 2.2 kΩ resistor (Ohm's law).

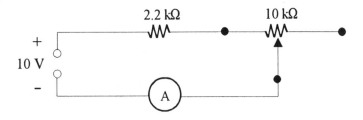

Figure 18-1

Minimum current: With the wiper at the extreme right position, the entire 10 kΩ resistance is included in series with the load. The current then is the source voltage divided by the sum of the load resistance plus the full rheostat resistance.

Materials

Power supply
Ammeter
10 kΩ potentiometer
One 2.2 kΩ resistor

Procedure

1. Before wiring the circuit, calculate the *highest current* the rheostat will allow.

2. Calculate the *lowest* current this circuit will allow to flow. _____

3. Wire the circuit as shown in Figure 18-1 above and Figure 18-2.

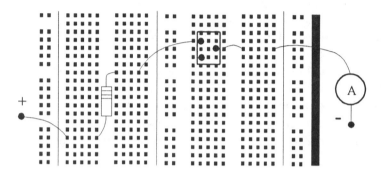

Figure 18-2

Do not apply power at this time. Set the rheostat to the center of its movement.

4. Switch the ammeter range to a scale that is *higher* than the maximum current you calculated above.

5. Turn the power supply control to zero position. Switch the supply on and slowly increase the voltage to 10 V. Watch the ammeter.

6. Is the current reading between your calculated maximum and minimum values? _____

7. Check the direction of operation. Does clockwise rotation cause current to increase or decrease? _____

8. Adjust rheostat for minimum current. Notice the ammeter. What do you measure for minimum current? _____

9. What was your calculated minimum current? _____

10. Adjust the rheostat for the highest current. What do you measure? _____

11. Record your calculated maximum current. _____

12. If you connect the 2.2 kΩ to the opposite end of the control, you can reverse the direction of operation.

    Refer to the schematic for Figure 18-1. Clockwise movement of the center terminal causes an *increase* in resistance to the circuit.

    Refer to the schematic for Figure 18-3. Clockwise movement of the wiper causes a *decrease* in resistance.

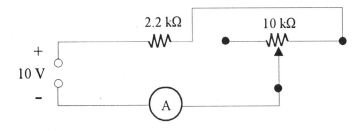

Figure 18-3

13. Rewire your circuit by connecting the 2.2 kΩ to the opposite end of the rheostat as shown in Figure 18-3.

14. Operate the control. Does the direction of rotation have the opposite effect on current flow?
    _____

19 Relays and Relay Circuits

Objectives

1. Identify the mechanical elements, contact forms, and electronic ratings of a typical electromagnetic relay.
2. Build and operate a relay switching circuit with output contacts switching a resistive load.
3. Design and build a special control circuit and draw the schematic or wiring diagram.

Preparation

Mechanical relays are manufactured in many different sizes, packages, contact configurations, and coil and contact ratings. They all have certain common characteristics, however, and the lab explores several of them. Relay functions are commonly defined by the contact arrangement and position of the armature, or moveable contact, when the relay is in the nonenergized state.

The specifications usually contain information for voltage rating of the coil and current/voltage limits for the contacts. Since this lab requires only a #40 incandescent bulb, almost all available relays should meet the requirement.

The following procedures will allow you to become familiar with the device and some relay circuit configurations.

Materials

One SPDT 12 V electromechanical relay
One #40 incandescent bulb with mounting socket
One volt-ohm-ammeter

Procedure

Part 1—Mechanical Configuration

Obtain the relay and find the specification data. This is usually printed on the side or top.

1. What is the voltage rating of the coil? _____ What voltage operates the solenoid? _____ Is it an AC or DC voltage? _____

2. What is the contact form? (SPDT, DPDT, etc.) _____

3. What is maximum current specified for the contacts? _____

   Is there a contact voltage indicated? _____ If so, what is the maximum contact voltage? _____

4. Visually inspect the relay. It may have a functional or pin diagram on one side. Find the two armature or coil pins. If the relay is open, trace the lead wires from the pins to the coil itself. If the device is enclosed, you will need to refer to the wiring or pin layout diagrams.

                        Initial here when you have found the coil connections: _____

5. Do the wires to the coil connect with any other wires or contacts on the relay? _____ Determine this visually or from the wiring diagram.

    If they do not, can there be any connection between the coil voltage and any of the contact circuits? _____

    Does this mean that the operating or coil voltage (the input) is separate and isolated from any of the contact circuits (the output)? _____

    This is, in fact, one of the principal uses of the relay: to isolate two circuits from each other.

    You can make any connection you want *outside* of the relay, but internally the operating coil and relay contacts are separate.

6. You have identified the coil pins, but prove this by measurement. Connect an ohmmeter to the solenoid pins and measure the coil resistance. If the value is in the hundreds of ohms, you have, in fact, found the coil. (It could be more or less depending on physical size.)

    Is there a value of coil resistance indicated in the specifications? _____ If so, how close is your reading to this value? _____

7. If the relay is open, trace the wiring for the moveable contact from its input pin, along the armature, to the contact. If the relay is enclosed, find the armature pin from the diagram.

                        Initial here when step 7 is done: _____

8. Next, find the *normally closed contact* (NC) input. This is the contact (and its associated connecting pin) that is in electrical contact with the armature.

9. You can prove that you have found the moveable contact and the normally closed contact by measuring the resistance between these two. Connect an ohmmeter between the terminal for the armature, or moveable contact, and the NC contact. It should measure zero ohms. This proves that you have found these two pins since they are the only terminals that will measure zero ohms in a properly working SPDT relay.

10. Find the *normally open contact* (NO). Measure the resistance between its terminal and the moveable contact terminal. This should measure *open* or "infinite" ohms. This measurement is not proof that you have the NO terminal since several other measurements can indicate open. The proof that you have found this will be in the next part.

Part 2—Relay Operation

In this section, you will energize the solenoid and observe the relay operation with electrical measurements.

1. Build the test circuit as required for your specific version of the SPDT relay. The diagram (Figure 19-1) can be used as a guide.

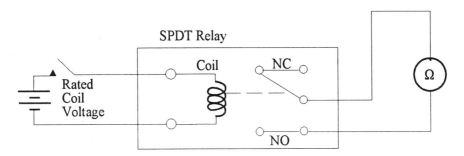

Figure 19-1 Test Circuit

2. Set the coil voltage to the rated value. Switch the voltage on and off several times. Listen for the sound of the armature operating and the make and break of the relay contacts. If the relay is not enclosed, try to observe the switching contact and the armature movement.

3. Switch the coil voltage off. Connect an ohmmeter to the NO set of contacts. The ohmmeter should read *open* or "infinite" with input voltage off. Switch on the input voltage and observe the ohmmeter. It should move to *zero* ohms, indicating the contacts have switched also. This demonstrates the action of these contacts and proves that you have, in fact, connected to the NO terminals.

   Switch off the input, and the ohmmeter should return to the *open* indication.

4. Connect ohmmeter to the NC contacts. These will read the opposite action: zero with input off and open with input on.

   Initial here when you have checked both contacts with ohmmeter: _____

Part 3—Relay Control Circuits

In this section you will build two different relay control circuits. The input voltage controls the position of the output contacts. These can be wired for direct or opposite action.
   Direct action means the output follows the input. When the input is on, output is on.
   Opposite action means: input on—output off.
   The first two circuits utilize the isolation feature of the relay circuit. Since the input is electrically and physically separate from the output, two separate voltage sources are used. One operates the input solenoid; the second is the power supply for the output circuit.

1. Look at the circuit in Figure 19-2. Is this circuit direct or opposite acting? _____

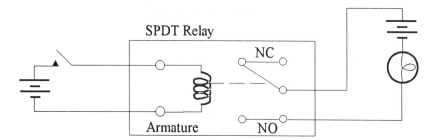

Figure 19-2 Relay Control Circuit

Test the bulb before depending on it for the circuit. Apply voltage directly to the lamp or measure its resistance. If it does not light or measures *open*, change it.

Be sure the circuit voltage is appropriate for the light bulb. You can set this supply voltage somewhat lower than the bulb rating to increase lamp life.

2. Operate the relay and confirm that the bulb is on when input is on and that the opposite occurs.

    Initial when you have the circuit working properly: _____

3. Move the negative lead to the opposite contact so that reverse-control action occurs.

    Initial when the circuit is working properly: _____

4. Figure 19-3 is a diagram that utilizes a single supply instead of the two-supply system. Input and output circuits are in parallel with the single source.

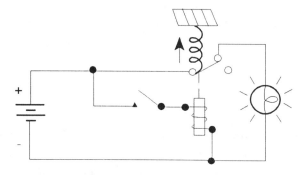

Figure 19-3 Relay Control Circuit

The advantage here is that only one power source is required. The relay contacts carry the possibly high load current. The operator switch only carries the relay coil current.

The disadvantage is that the isolation feature is no longer in effect, and both input and output must utilize the same voltage.

5. Wire Figure 19-3 with a single supply and operate the relay. You will achieve opposite action with this circuit.

Initial when you have completed and operated this circuit: _____

Part 4—Special Project

Design and build a relay control circuit utilizing two voltage sources and two incandescent #40 (or similar) bulbs.

Connect the circuit in such a way that when the relay solenoid is energized bulb A is on and bulb B is off. When the relay solenoid is *de*-energized or turned off, the opposite occurs and the lights switch: Bulb A is off; bulb B is on.

If they are both on or both off at the same time, or if only one bulb lights, the circuit is incorrect.

Be sure to check that each bulb is working correctly before building your circuit.

To help you in designing the circuit, begin by drawing a wiring diagram. When you have made any changes necessary to make the circuit work, draw a schematic diagram, a "clean" version of the wiring diagram, and include these as part of the project.

Initial here when project is complete and working: _____

Instructor initial here (optional): _____

Lab Questions

1. What part of the relay is considered the input?

2. What section of the relay is considered the output?

3. Is there an electrical connection between the input and output of the device?

4. What DC resistance would be measured between the normally closed contact and the coil of the relay?

5. What other electronic device is physically isolated between input and output?

20 Internal Resistance and the Loaded Power Source

Objectives

1. Build a simulated or equivalent nonregulated voltage source.
2. Observe the effect of load current on the output.
3. Verify the relationship of load current to output voltage.

Preparation

The experiment shows the effect of various load currents on the output of a simulated voltage source.

Refer to the circuit of Figure 20-1. The internal generator voltage is indicated by V_S. The internal generator resistance is indicated by R_S.

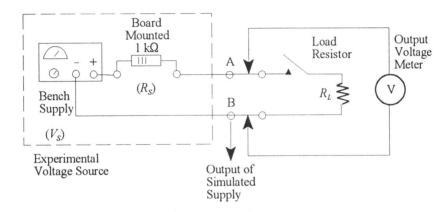

Figure 20-1

For this project, you will use your laboratory supply to simulate the internal voltage.

The source resistance is simulated by the 1 kΩ board-mounted resistor, R_S. Although R_S is mounted on the breadboard, it is considered a part of the supply circuit for this project.

Points A and B represent the output terminals of the simulated supply.

R_L is the load or external resistance. As you change the external resistance, you will be changing the amount of current delivered by the supply. As the current changes, the I_R drop across the internal resistance will change and, therefore, the output voltage as well.

Look at the placement of the voltmeter. It is on the "supply" side of the wire switch. It must be in this location for you to observe the difference between loaded and unloaded output.

Materials

Regulated DC voltage supply
One VOM meter
Resistors: 1/2 watt
 One 100 kΩ, one 47 kΩ, one 10 kΩ, one 5.6 kΩ, one 2.2 kΩ, two 1 kΩ

Procedure

1. Build the circuit as shown in Figure 20-1. Make a mental separation of the part of the circuit that represents the inside of the power source and the components that represent the external circuit.

    Set the internal source voltage with the same meter you will be using to measure load voltage. Set it to exactly 10 V.

2. Look at Table 20-1. The column headings indicate the steps to be followed in obtaining information.

    Column 2: Begin with a 100 kΩ resistor for the load. Open the switch and measure the output voltage without a connected load. This is V "no load" or V_{NL}.

    It should read 10 V with no current flowing from the supply. If it reads zero, it probably means your meter is on the wrong side of the switch.

    Measure V_{NL}. Enter in column 2. _____

3. Column 3: Close the switch and measure output voltage again. You should see a small drop in voltage due to the current flowing through the 1 kΩ resistor.

    Open and close the switch several times to identify the measurement.

    Measure V_L with 100 kΩ load. Enter in column 3. _____

4. Column 4: Calculate the voltage that should appear across the 100 kΩ load resistor when you close the switch. _____

    Note: This would be a good place to use the voltage divider equation.

    V_L calculated is: $V_L = V_S \times \dfrac{R_L}{R_S + R_L}$

5. Column 5: Calculate current in the circuit under load. _____

    Calculated current is: $I_L = \dfrac{V_S}{R_S + R_L} = \dfrac{V_L}{R_L}$

6. Open the switch. Change the load resistance to the next value indicated in the table.

7. Repeat the calculations and measurements. The drop in voltage will be more apparent with the 47 kΩ load.

8. Complete the data table by making measurements and calculations as indicated.

Table 20-1 Data Table for Figure 20-1

1	2	3	4	5
Load Resistance	Measure Output Voltage No Load	Measure Output Voltage With Load	Calculate Output Voltage With Load	Calculate Load Current
100 kΩ				
47 kΩ				
10 kΩ				
5.6 kΩ				
2.2 kΩ				
1 kΩ				

Lab Questions

1. Did the largest or smallest load resistor cause the most current to flow? _____

   Does this conform with Ohm's law? _____

2. What happened to the output voltage as the load current increased?

3. In your own words, explain why output voltage dropped when output current increased.

4. Regulated supplies, such as the type found in most school labs, maintain their output constant over a range of load currents.

   The equivalent circuit for these supplies sometimes shows a battery in series with a very small internal series resistor, such as 0.1 Ω.

   Explain why this equivalent circuit would not show a significant drop in output voltage as load current increased.

5. Compare the regulated circuit of step 4 with the equivalent circuit of this experiment.

6. The ability of a voltage source to maintain constant output is sometimes defined as the percent of regulation. It is indicated by the formula:

$$\% \text{ Regulation} = \frac{V_{\text{No Load}} - V_{\text{Load}}}{V_{\text{Load}}} \times 100\%$$

   Calculate the percent regulation for your simulated supply with the 10 kΩ load resistor. Use the data from the experiment.

21 Maximum Power Transfer Theorem

Objectives

1. Find by experiment the load resistance value that generates the maximum power from an imperfect voltage source.
2. Plot a graph of load resistance versus load power.
3. Determine the circuit power efficiency for different loads.

Preparation

The project lets you see the principles of maximum power transfer by observing the changes in load power as load resistance changes. The efficiency of the power source is derived from circuit measurements.

R_S represents the internal resistance of a nonregulated voltage source.

V_S is the voltage generator and is assumed constant. For the experiment, your V_S should be a regulated supply. V_S in series with source resistance represents the complete "simulated" supply.

R_L is the load resistance.

The graph is plotted with R_L on the horizontal axis and load power on the vertical axis. It would be helpful to indicate the circuit values and perhaps add a schematic somewhere on the graph paper. This helps identify the circuit being depicted by the graph data.

Materials

Regulated DC voltage supply
One VOM meter
Resistors: 1/2 watt
    One 1 kΩ, one 2.2 kΩ, one 3.3 kΩ, two 5.6 kΩ, two 10 kΩ, one 22 kΩ, one 47 kΩ

Procedure

1. Refer to the circuit of Figure 21-1. The dotted outline represents a nonregulated power supply with internal voltage generator V_S and internal resistance R_S. The load resistor is considered external to the supply.

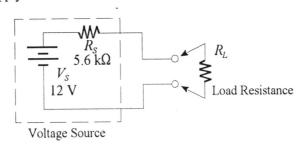

Figure 21-1

Build the simulated supply with 1 kΩ load.

2. Measure voltage across R_L. $V_L =$ _____

3. Calculate circuit current from V_L. $I =$ _____

$$I = \frac{V_L}{R_L}$$

4. Calculate load power for the 1 kΩ. $P_L =$ _____

$$P_L = I \times V_L$$

5. Calculate the total circuit power. $P_T =$ _____

$$P_T = I \times V_T$$

6. Find the power transfer efficiency. $P_{eff} =$ _____

$$P_{eff} = \frac{P_L}{P_T} \times 100\%$$

7. Enter each of the above numbers into Table 21–1 for the 1 kΩ load resistance.

8. Complete the table for each load resistance.

9. You can demonstrate that the principle of maximum power transfer is true for *any* value of source resistance.

   Change R_S to 10 kΩ and determine the values for Table 21–2.

10. Using the graph paper at the end of this experiment or your own graph paper, plot a graph of R_L versus P_L for the measured load voltage from Table 21–1.

Table 21-1 $R_S = 5.6$ kΩ

Load Resistance R_L	Measure Load Volts V_L	Calculate Current I	Calculate Load Power P_L	Calculate Efficiency P_{eff}
1 kΩ				
2.2 kΩ				
3.3 kΩ				
5.6 kΩ				
10 kΩ				
22 kΩ				
47 kΩ				

Table 21-2 $R_S = 10$ kΩ

Load Resistance R_L	Measure Load Volts V_L	Calculate Current I	Calculate Load Power P_L	Calculate Efficiency P_{eff}
1 kΩ				
2.2 kΩ				
3.3 kΩ				
5.6 kΩ				
10 kΩ				
22 kΩ				
47 kΩ				

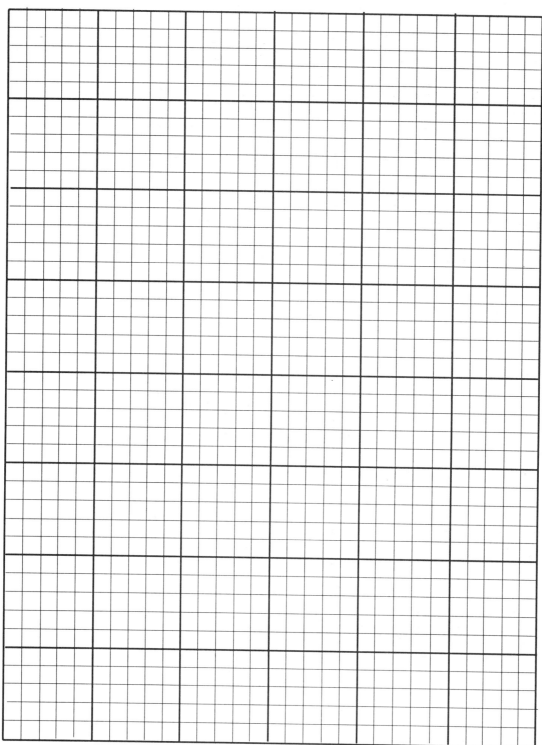

Power P_L

Load Resistor R_L

Lab Questions

1. In Table 21-1, when did maximum power occur in the load resistance?

2. In Table 21-2, R_S was changed to 10 kΩ. What happened to the value of load resistance required to develop the most power?

3. Did the highest circuit efficiency occur when R_L was low, when R_L was high, or when $R_L = R_S$?

4. Circuit efficiency refers to the amount of power developed by the load compared to the total circuit power.

   Using this definition, explain why the efficiency was worse with the 1 kΩ load compared to a 5.6 kΩ load.

5. Describe an application where obtaining the most power from a source is important.

22 Thevenin's Theorem

Objectives

1. Reduce a complex circuit to its Thevenin equivalent by calculation and laboratory measurement.
2. Verify the equivalence of the two circuits.

Preparation

The experiment presents the Thevenin equivalent circuit in two ways:
   Part 1 steps through the calculation of circuit reduction to review the procedure.
   Part 2 uses a breadboard circuit and measurement to find the Thevenin components.

Materials

DC power supply
Analog or digital VOM
Resistors: 1/2 watt
   One 1 kΩ, one 2.2 kΩ, two 3.3 kΩ, three 5.6 kΩ, four 10 kΩ

Procedure

Part 1—Calculations

1. Refer to Figure 22-1. This is the original series-parallel circuit. It represents a voltage source supplying a load resistance at terminals A and B.

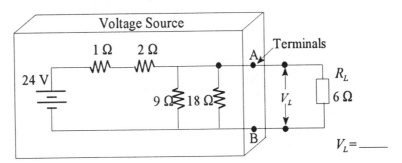

Figure 22-1

Assume a 6 Ω load resistance is connected at A and B. Using previous methods, calculate the voltage that will appear across this resistor. The values have been chosen to make the solution work out "neatly" with whole numbers.

   Calculate V_L for the circuit in Figure 22-1 = _____

The next steps are the calculations that convert the circuit to its Thevenin equivalent.

103

2. Calculate the open circuit voltage that occurs at points A-B with R_L removed. This is the Thevenin voltage V_{Th} (Figure 22-2).

   Voltage between A and B with R_L removed: V_{Th} = _____

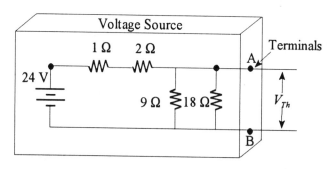

Figure 22-2

3. Short the original voltage source, or remove it, and replace it with a wire. Calculate the resistance between A and B with the battery shorted. This is the Thevenin equivalent resistance, or R_{Th} (Figure 22-3).

   Resistance between A and B with V shorted: R_{Th} = _____

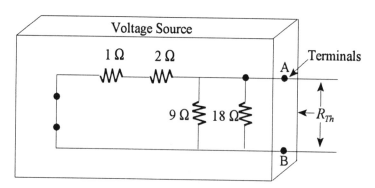

Figure 22-3

4. Your calculation should result in: $V_{Th} = 16$ V, $R_{Th} = 2\ \Omega$

   The equivalent circuit is V_{Th} in series with R_{Th} (Figure 22-4).

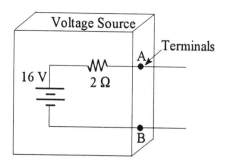

Figure 22-4

5. Connect the 6 Ω resistor to terminals A-B and calculate the voltage across the load resistor (Figure 22-5).

   A simple voltage divider equation should do the job.

   Calculate voltage A-B = _____

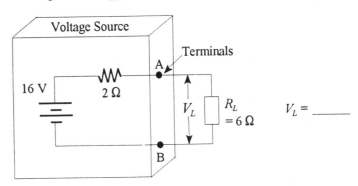

$V_L =$ _____

Figure 22-5

The voltage is the same as the original circuit.

You can replace R_L with any value and use the equivalent circuit to find the output voltage and current.

Part 2—Finding Thevenin Equivalent Circuit by Measurement

Complex Circuit

1. Build the series-parallel circuit of Figure 22-6. This is the circuit you will simplify.

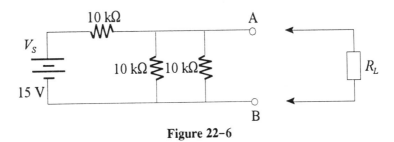

Figure 22-6

2. Connect a 1 kΩ load resistor to points A-B. Measure the voltage across this resistor and enter in Table 22-1. Connect the other resistors and measure their voltage. Complete the table for the original circuit.

Thevenin Equivalent Circuit

3. Remove the load resistor.

   Measure the voltage between A and B with no load. This is the Thevenin voltage (Figure 22-7).

   $V_{Th} =$ _____

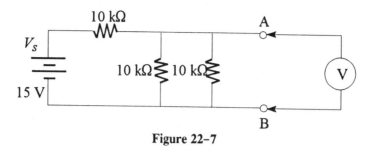

Figure 22-7

4. Switch off or remove the voltage source. Connect a shorting wire between the + and - leads or terminals.

   Measure the resistance between A and B. This is Thevenin resistance (Figure 22-8).

   R_{Th} = _____

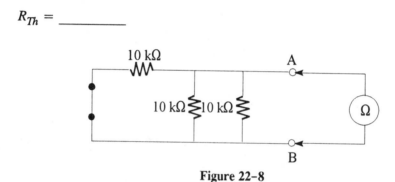

Figure 22-8

5. Build the equivalent circuit consisting of the voltage, V_{Th} in series with the resistance R_{Th}.

   This is the Thevenin equivalent circuit (Figure 22-9).

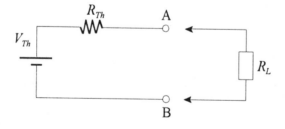

Figure 22-9

6. Now prove that the two circuits are equivalent. Connect the load resistors of Table 22-1 to the Thevenin circuit and measure the voltage across each one. Compare the results of the two circuits.

Table 22-1 Data to Compare Equivalent Circuits

Load Resistor	Measure V_L with Original Circuit	Measure V_L with Thevenin Circuit	Compare Voltages
1 kΩ			
2.2 kΩ			
3.3 kΩ			
5.6 kΩ			
10 kΩ			

Lab Questions

1. In your own words, explain why the voltages measured with the Thevenin equivalent circuit are the same as the voltages obtained with the original circuit.

2. According to the maximum power transfer theorem (Experiment 21), the most power is delivered to a load when the load resistance is the same value as the internal resistance of the power supply.

   Which of the resistances in Table 22-1 would deliver the maximum power to the complex circuit? Why?

3. Would a load resistor have the same current in both the original and Thevenin circuits? Why?

4. Would a load resistor develop the same power in both the original and Thevenin circuits? Why?

5. Does the battery of the original circuit and the battery of the equivalent circuit deliver the same *total* circuit power?

23 The Wheatstone Bridge

Objectives

1. Build a Wheatstone bridge using three fixed resistors and a potentiometer.
2. Adjust the bridge to achieve a null or balanced condition.
3. Make appropriate calculations to find the value of an unknown resistance and confirm the operation of the bridge.

Preparation

Refer to the schematic and the suggested wiring diagram, Figures 23-1 and 23-2.
 You can use a different board layout; but if you place the circuit components in the same general arrangement as the schematic, you may find that measurements and troubleshooting are more understandable.
 R_X is the unknown resistance; R_V is the variable resistance.
 A laboratory quality bridge would have a dial plate that accurately reads the resistance setting. In this case, you will use a PC potentiometer and check its resistance with an ohmmeter.
 Meter "G" is a galvanometer with center zero reading. If this is not available, a practical substitution would be a standard analog voltmeter.
 The experiment assumes you will be using a voltmeter. You can create a sense of galvanometer display by mechanically adjusting the pointer so that it reads "up" scale. Whichever mark it indicates with no voltage applied is the "zero" indication.
 Remember, you are primarily interested in observing null, or zero, voltage. If you know the meter mark that represents zero, you can identify zero voltage condition.
 Be sure to readjust the meter to a proper mechanical zero before taking any normal voltage measurements.

Materials

Ohmmeter
Analog voltmeter or galvanometer
10 kΩ PC potentiometer
Resistors: 1/2 watt
    One 2.2 kΩ, two 1 kΩ, one 4.7 kΩ

Procedure

Part 1—Bridge Operation

1. Build the circuit with power *off* using the wiring diagram (Figure 23-2).

2. Select a voltmeter range which is high enough to include the value of source voltage. Next, turn on the source voltage.

3. Adjust the potentiometer so that the meter reads "up" scale.

4. Slowly adjust the variable resistor (potentiometer) until the meter reading moves to zero. (This is the procedure called "balancing" the bridge.)

5. Next, select a lower or lowest scale. You may find that a reading occurs. (The high scale was not sensitive enough to show this small voltage.)

   Slowly adjust the variable resistor until zero reading is again obtained.

6. Switch voltmeter to a higher scale to protect it from overranging when you change the circuit.

7. Now measure the resistance of R_V with an ohmmeter.

   Suggested Technique: Switch off the source voltage. Leave the potentiometer connected (plugged in) to the board but disconnect the upper lead of R_X. Measure the resistance of R_V by placing the ohmmeter leads directly on the connecting wires of the potentiometer.

8. Calculate R_X. _____ Is it reasonably close to the color code value?

   $$R_X = \frac{R_1}{R_2} \times R_V = \underline{\qquad}$$

9. Change R_1 to 1 kΩ. Balance the bridge again and measure the new value of R_V. _____

   Calculate R_X. _____ Is it still about 4.7 kΩ? _____

10. (Optional) If your lab has a set of unknown resistances with values between 1 kΩ and 9 kΩ, select one of these and measure using the above procedure.

Part 2—Circuit Analysis

1. According to circuit theory, should the voltage ratios on each side of the bridge be the same when the bridge is balanced? _____

   Balance the bridge and measure with a voltmeter:

   $V_1 = \underline{\qquad}$

   $V_2 = \underline{\qquad}$

   $V_X = \underline{\qquad}$

   Ratio $\dfrac{V_1}{V_2} = \underline{\qquad}$

   $V_V = \underline{\qquad}$

   Ratio $\dfrac{V_X}{V_V} = \underline{\qquad}$

2. Are the voltage ratios of the two sides the same? _____

3. When the bridge is balanced, is the voltage from point A to ground the same as the voltage from point B to ground?

4. With the bridge operating and balanced, measure A to ground. _____ Measure B to ground. _____

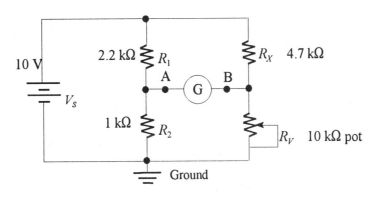

Figure 23-1 Wheatstone Bridge Schematic

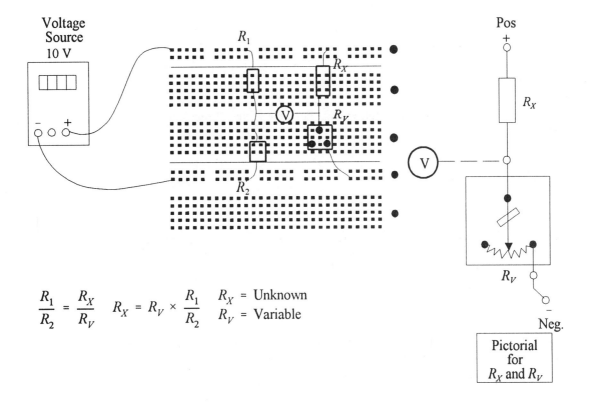

$$\frac{R_1}{R_2} = \frac{R_X}{R_V} \quad R_X = R_V \times \frac{R_1}{R_2} \quad \begin{array}{l} R_X = \text{Unknown} \\ R_V = \text{Variable} \end{array}$$

Figure 23-2 Wheatstone Bridge Wiring Diagram

Lab Questions

1. When the bridge is balanced, the left and right side ratios are equal. $\dfrac{R_1}{R_2}$ is equal to $\dfrac{R_X}{R_V}$.

   Are the top and bottom ratios equal as well? _____ Does $\dfrac{R_1}{R_X} = \dfrac{R_2}{R_V}$? _____

2. Prove your answer to question 1. Set up the bridge with applied voltage in null or balanced condition.

   Measure V_1. _____ Measure V_X. _____

   Measure V_2. _____ Measure V_V. _____

   Calculate the ratio of $\dfrac{V_1}{V_X}$. _____

   Calculate the ratio of $\dfrac{V_2}{V_V}$. _____

   Is the top arm ratio equal to the bottom arm ratio? _____

3. In your own words, explain why the voltage between A and B reads zero when the bridge is balanced.

4. Why does the galvanometer (or voltmeter used as galvanometer) sometimes indicate positive voltage and sometimes negative voltage?

5. When the bridge is near balance condition, a very small voltage occurs across the null meter. When the bridge is out of balance, or the arm ratios are very different, a voltage near the source voltage can occur across the meter.

   Describe an experimental method to protect the sensitive galvanometer while the bridge is being adjusted.

24 Oscilloscope and Waveform Generator: Familiarization

Objectives

1. Observe the operation of a waveform generator and oscilloscope.
2. Adjust the controls to learn their respective functions.
3. Measure DC voltage with the oscilloscope.

Preparation

The lab allows you to observe the patterns generated by an electronic oscillator (waveform or signal generator). You will learn how to adjust the generator and oscilloscope controls to create and observe different voltage patterns.

The lab begins with an oscilloscope/generator combination presenting a 5 kHz sine wave. You will adjust the generator controls, one at a time, and observe the effect of each adjustment on the pattern.

You then adjust the oscilloscope controls to determine their effect on the pattern presentation. The initial setup must exist before the project can begin. If you are having difficulty obtaining the pattern, ask for help.

Materials

Frequency generator
Oscilloscope

Equipment Setup

Connect the output of the generator to one input of the oscilloscope (Figure 24-1). The oscilloscope should present about 4 cycles of a 5 kHz sine wave, 2 V_{p-p}, covering 4 vertical divisions (Figure 24-2). The following control settings should create the pattern.

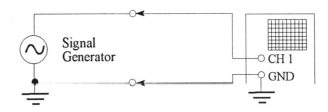

Figure 24-1

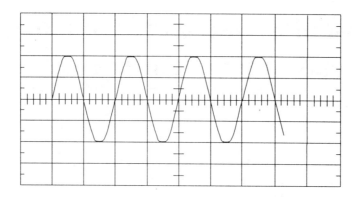

Figure 24–2

Oscilloscope
VOLTS/DIV:	0.5 V/div
Sweep Frequency:	0.1 ms
VERNIER or VAR:	Calibrate or CAL
AC-GND-DC:	AC
Triggering:	Auto
Level:	Zero or Centered
Slope:	Positive
Vertical adjust:	Center on screen
Horizontal adjust:	Center on Screen

Generator
Frequency:	5 kHz
Amplitude:	2 V_{p-p}

If you cannot obtain this initial, starting presentation, ask your instructor to create these conditions for you.

Note: After obtaining the initial conditions, it is important that you do not make any adjustments unless called for in the procedure. In time you will be free to adjust any controls and make your own decisions on adjustment.

At this point in the course, however, things will progress more smoothly if you follow the instructions in the order written.

Procedure

Part 1—The Waveform Generator

Amplitude or Output

1. Find the control marked amplitude, level, or output. This is the adjustment that changes the generator's output voltage. Notice the number of vertical spaces or divisions covered by the pattern on the oscilloscope.

2. Slowly turn the generator amplitude control to the right and left and observe the pattern on the screen.

3. Adjust the generator output to return the display to its original size, 4 vertical divisions peak-to-peak.

4. This can also be considered the "volume" control for the generator. If you could listen to this sine wave on a loud speaker, you would hear the sound getting louder and softer as you adjust the amplitude. The oscilloscope presents this information by the size of the pattern on the screen.

Frequency

1. Find the frequency adjustment and notice the manner in which it indicates the output frequency. Are the units in hertz or kilohertz? _____ Slowly increase the frequency and notice the effect on the oscilloscope display. What happens to the pattern? _____ Are there more or fewer cycles displayed when you increase the frequency? _____

2. Slowly lower the frequency output and notice the effect on the screen. Are there more or fewer cycles displayed when you lower the frequency? _____

3. Return the frequency control to 5 kHz.

Pattern

1. Most signal generators have three different waveform patterns available at the output. Find the pattern selector switch. It should be set to produce a sine wave pattern.

2. Move the selector switch to the sawtooth pattern. Notice the oscilloscope presentation. Has the amplitude or vertical height of the pattern changed? _____ Has the number of cycles on the screen changed? _____

3. Select the square wave pattern. Notice the scope presentation. Did the height of the display change when you changed the pattern? _____ Did the number of cycles on the screen change? _____

4. Return the pattern selector switch to the sine wave position.

Part 2—Oscilloscope Controls

On-Off Switch

The power switch may be a part of the intensity control or may be separate. Once switched on, the scope may be left on for the duration of the experiment. It is not necessary to turn it off after each measurement.

Intensity

Adjust the control to a comfortable viewing brightness. If a glow exists around or alongside the pattern, the intensity is too high and should be reduced.

Focus

Adjust the focus control for maximum sharpness of the trace.

Horizontal Position

Slowly move the horizontal control. Move it left and right and observe the effect. Adjust the position of the pattern so that it begins one division in from the left side of the grid. The setting is for purposes of this experiment only. The pattern can begin at the left edge when you are performing experiments.

Vertical Position

Move the vertical position up and down and finally place the display at the center of the screen.

AC-GND-DC *Mode*

1. Move the switch to GND position. The trace should become a straight horizontal line.

   The GND position connects the measuring terminal to ground and removes the input voltage from the display. Since ground is the zero reference for measurements, the trace position represents *zero* voltage. The horizontal line is sometimes called the reference, zero, or base line.

2. Is the reference line at the exact center of the screen? If it is not, adjust the vertical position so that the line overlaps the center X axis.

Trace Rotation

1. Is the reference line exactly horizontal, or does it slant one or two degrees?

2. If your oscilloscope has an adjustment called *Trace Rotation*, turn it slowly both ways and notice the effect on the ground reference line.

3. Adjust the trace so that it is horizontal.

4. Return AC-GND-DC switch to AC position. The sine wave should be displayed once again.

Sweep Trigger Controls

Slope

1. The sweep circuit can begin tracing the sine wave when the voltage is becoming more positive or when it is becoming more negative. Find the slope switch. It should be set to POS. (Some scopes have a small slanted mark to indicate the direction of change.) Look at the start of the sweep. Is the pattern moving in a negative direction or in a positive direction? _____

2. Move the slope switch to NEG. What happened to the waveform? _____ Is the pattern now moving in a negative or positive direction? _____

3. Move the slope back to POS.

Level

This control determines the voltage level at which the trace will begin painting the waveform. If the control is set at zero or center, the sine wave trace will begin the trace when the waveform is about zero volts. This particular control does not have calibrated markings since the setting is usually determined by observation. The word *level* refers to the voltage level or position at which the trace begins.

1. *Slowly* rotate the level control and watch the starting position of the waveform.

2. Move it positive and then negative. Notice what happens to the starting position of the pattern.

3. What happens to the pattern when you move the level to a position higher or lower than the sine wave itself?

4. Move the level back to zero.

Source

AUTO: When the trigger level is compared with the signal applied to the input terminals, the triggering is said to be automatic. The "source" of the sweep trigger level is then the input voltage. This is the AUTO position.

EXT: When the trigger level is compared with a voltage applied to a separate terminal, not the input, the triggering is said to be coming from an "external source." A special terminal or jack is usually supplied for this.

LINE: In this position, the level is compared with the AC power line voltage.

Volts/Division **(VOLTS/DIV)**

This control is the vertical sensitivity adjustment for the input amplifier. It allows you to measure the amplitude of the waveform presented on the screen. It also acts as a magnifier or zoom-in, zoom-out. It does not actually change the value of the applied voltage. This input can only be changed by the signal generator.

Variable (VAR): This control may also be labeled VERNIER or FINE. It is commonly mounted concentrically on the same shaft or next to the VOLTS/DIV switch. The VAR adjustment changes the calibration or value of the VOLTS/DIV setting. It must be set on its CAL or calibrate position for the switch settings to correctly indicate the values labeled. Most of your experiments, including this one, will be done with the VAR adjustment in its calibrate position.

1. Place the variable adjustment for the vertical sensitivity (VAR) into its calibrate position. It may have an indent or "click" at this location.

2. Adjust the VOLTS/DIV setting one position clockwise. What happened to the size of the waveform?

3. Adjust the VOLTS/DIV setting one position counterclockwise from its original position. What happened to the size of the waveform?

4. Return the voltage sensitivity control to its original position (0.5 V/div).

5. *Measure* the peak-to-peak value (p-p) of the waveform as follows. Count the number of squares, centimeters, or divisions covered by the waveform—top to bottom or peak to peak. Multiply this number by the 0.5 V/div setting of the control. Record these data here.

   VOLTS/DIV setting = _____ Number of divisions = _____

   Voltage Peak to Peak = _____

6. Change the voltage sensitivity to 1 V/div. Measure the peak-to-peak value again. Does it still measure the same? Did changing the sensitivity control change the actual value of the input signal?

7. Return sensitivity to 0.5 V/div.

   Increase the input amplitude until the picture covers 6 divisions top to bottom. Measure the peak to peak of this pattern. Is the value greater than step 4? Did adjusting the generator amplitude actually change the voltage of the input signal?

Time/Division (TIME/DIV)
The TIME/DIV control allows you to measure the frequency and period of the input waveform. It determines the amount of time represented by each division of the horizontal axis. What is the setting of the control now? _____
   Adjusting this control does not change the input frequency. It only expands or contracts the time base and, therefore, the number of cycles appearing on the screen. In effect, it is a zoom-in or zoom-out for the horizontal display.
   VAR/VERNIER or FINE: This control has the same effect on the TIME/DIV calibration as the corresponding control for vertical sensitivity discussed previously.

1. Check that the variable adjustment is in its calibrate position.

2. Move the TIME/DIV control one position. What happened to the number of cycles appearing on the screen?

3. Move the control to its original position. What happened to the number of cycles appearing on the screen?

   Have you changed the frequency setting on the signal generator? _____ If you did not change the generator setting, did the actual input frequency change when you adjusted the time base? _____

4. Slowly increase the frequency setting of the waveform generator. What happened to the number of cycles appearing on the screen?

   Since you did not change the oscilloscope time base, does the greater number of cycles appearing on the screen represent a higher frequency? _____

5. Slowly decrease the frequency setting to its original value, 1 kHz. What happened to the number of cycles appearing on the screen?

Part 3—DC Measurements

1. The oscilloscope can be used to indicate DC voltages as well as AC patterns. Disconnect the AC generator and connect a variable DC voltage source to the scope input.

    Check that grounds are common, DC power is off, and voltage control is turned to zero.

2. The scope frequency setting is not important if you are measuring DC, but the sweep should be fast enough that a solid horizontal line appears on the screen.

3. Establish the zero volts position of the reference line by placing the mode switch on GND. Any location of the line with input at GND represents zero input voltage.

    Move the vertical placement so that reference is one major division above bottom of graticule.

4. Move the VOLTS/DIV switch to 1 V/div. Be sure that the calibration knob is on calibrate position.

5. Switch on the DC voltage source and slowly increase the input voltage until the horizontal line is 4 divisions above zero reference.

6. Since each division represents one volt and the DC input moved the indicator line 4 divisions, the scope is now reading 4 V DC.

    Measure the DC input with a DC meter. Does the DC meter confirm your reading with the oscilloscope? _____

7. Move the VOLTS/DIV setting to 2 V/div. What happened to the deflection?

    Multiply the number of divisions above zero reference by the 2 V/div calibration, and you should still be reading 4 V DC.

8. With VOLTS/DIV on 2, increase the input until scope indicates 4 divisions. What is the DC voltage? _____ Check your reading with DC meter.

9. Switch the mode selector to GND. Move the zero reference to a position 2 divisions above the bottom of the graticule.

10. Move selector to DC and read input voltage. Does the position of the reference affect the reading? _____

Lab Questions

1. What factors might cause the DC reading of the oscilloscope and the DC meter to be different?

2. Is the position of the ground or zero reference line important to the value of voltage indicated?

3. Assume zero reference is at center of screen. If the horizontal line moves down, what would this indicate about the polarity of the DC voltage?

4. Does your scope have a polarity reversal switch similar to the DC meter?

5. A technician is measuring an AC waveform and notices that the top part of the pattern is lost above the top of the graticule. Which control should be adjusted to bring the waveform to the center of the screen?

6. A technician is measuring an AC waveform. When an adjustment is made to the circuit, the waveform becomes smaller and then breaks into multiple moving sine waves. Which control should be adjusted to stop the movement?

7. A solid line appears on the screen and no input, AC or DC, will move the line. Which control should be moved to bring the pattern on screen?

8. A white haze occurs on both sides of the trace. What is happening, and what adjustment should be made?

9. The screen has no pattern or reference line at all. Name several controls that may be causing the problem and list the adjustments you would make?

10. What would be the effect on an AC waveform if the trigger input switch is in the wrong position?

25 Measuring Time and Frequency with the Oscilloscope

Objectives

1. Measure time and calculate frequency with the oscilloscope.
2. Perform several measurements to develop a degree of skill in reading the oscilloscope.
3. Use the scope pattern to set the input waveform at a predetermined frequency.

Preparation

The lab provides practice in making waveform measurements with the oscilloscope. The data table allows you to see the approximate accuracy of your measurements. If your readings are incorrect, you will be aware of this and can check your work.

Skill comes from many repetitions of a procedure. You will not become expert after this one project, but there are enough different readings for you to become reasonably skillful in performing these measurements.

Materials

Frequency generator
Oscilloscope

Procedure

Connect the output of the signal generator to Channel 1 of the oscilloscope. Connect high to high and ground to ground. This is the same connection used in Experiment 24.

Generator Setting

Adjust the oscillator frequency for a 5000 Hz sine wave. The amplitude or voltage can be any convenient value.

Oscilloscope Settings

Use the basic settings from Experiment 24.

Oscilloscope

VOLTS/DIV:	0.5 V/div
Sweep Frequency:	0.1 ms
VERNIER or VAR:	Calibrate or CAL
AC-GND-DC:	AC
Triggering:	Auto
Level:	Zero or Centered
Slope:	Positive
Vertical adjust:	Center on screen
Horizontal adjust:	Center on Screen

1. Adjust the voltage sensitivity (VOLTS/DIV control) so that the sine wave covers 6 vertical divisions peak-peak. Some instruments label this adjustment VOLTS/CM.

2. Adjust the sweep frequency (TIME/DIV control) to create 2 or 3 cycles. Some instruments label this adjustment TIME/CM. Record the setting here. _____

3. Make sure the TIME/DIV calibration adjustment is in the CAL position.

4. Adjust the vertical position so that the pattern is centered on the screen.

Time and Period Measurement

The time for one cycle of the waveform is measured by counting the number of divisions occupied by one cycle and multiplying by the TIME/DIV setting (Figure 25-1).

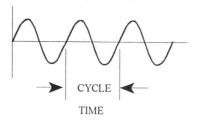

Figure 25-1

1. Select one cycle from the pattern you would like to measure. Adjust the horizontal position so that the pattern cuts across the center of the screen at a major division mark.

2. Count the number of divisions from this point to the same point on the next wave. Be sure to include tenths of a division.

3. Multiply the number of divisions by the TIME/DIV setting. Record this value in column 2 of Table 25-1 for 1 kHz. Note: Time units should be in milliseconds for a frequency in kilohertz.

Frequency Calculation

1. Calculate the reciprocal of the time and enter this in column 3 as the frequency. Units should be kilohertz.

2. Compare the difference between the frequency calculated in column 3 and the frequency entered in column 1. If these numbers are the same, it indicates you have read the oscilloscope correctly for this measurement.

3. Complete columns 2, 3, and 4 for the remainder of the frequencies listed.

4. Change the generator pattern from sine to square wave.

5. Measure time for one repetition of the square wave, at each of the frequencies listed. Refer to Figure 25-2. Compare this reading with the value in column 2.

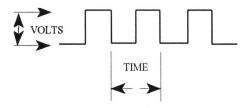

Figure 25-2

Table 25-1 Data for Figure 25-1 and Figure 25-2

1	2	3	4	5	6
Set Frequency Generator	Measure Period in Milliseconds	Calculate Frequency in Kilohertz	Compare Columns 1 and 3	Measure Time for Square Wave	Compare Columns 5 and 2
1 kHz					
2 kHz					
4 kHz					
5 kHz					
10 kHz					
15 kHz					
20 kHz					
40 kHz					
100 kHz					

Setting Frequency with the Oscilloscope

The oscilloscope can be used as the measurement standard when adjusting the frequency of the incoming signal. In this situation, you adjust the frequency of the oscillator so that a certain display occurs on the screen.

Example: To set the incoming signal for 5 kHz.

1. Determine the number of cycles you would like displayed on the screen or the number of divisions you want for a single waveform. (Use 4 divisions for this example (Figure 25-3.))
   _____

2. Find the time for one period by calculating the reciprocal of the frequency. (In this example: 1/5 kHz = 0.2 ms.) _____

3. Divide the 0.2 ms by the 4 divisions of one period to find the time for one division (0.2 ms/4 = 0.05 ms/div).

4. Set the sweep frequency or TIME/DIV switch for 0.05 ms. Be sure the variable adjust control is on calibrate.

5. Adjust the signal generator to approximately 5 kHz. Make small adjustments with the frequency control until the waveform pattern covers exactly 4 divisions. (Move the horizontal position to assist in reading the display.)

6. Compare the "reading" obtained with the scope to the frequency setting on the oscillator. They may be slightly different. Oscilloscopes are often calibrated more accurately.

7. Repeat the procedure with a different frequency, such as 4 kHz. Compare your results from the scope with the frequency indication of the signal generator.

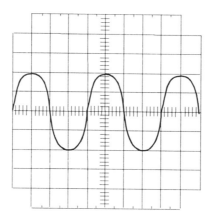

Figure 25-3

Lab Questions

1. If the time or sweep *calibrate* control is not at its CAL position, will the time reading be long or short? _____

   Will the frequency calculation be high or low? _____

2. If the VOLT/DIV control is changed, will this affect the reading of frequency? _____

3. While measuring the frequency output of a circuit, a technician notices that the number of sine wave cycles on the screen has increased. No adjustments have been made to any of the oscilloscope controls.

   Has the frequency increased, decreased, or remained the same? _____

4. When you adjust the TIME/DIV to a lower setting, you notice this causes more cycles to be displayed on the screen.

   Has the input frequency increased, decreased, or remained the same? _____

5. While working on an AC circuit, you notice that the amplitude or peak-to-peak voltage of the display has increased. The number of cycles displayed is the same.

   Has the input frequency increased, decreased, or stayed the same? _____

26 Measuring AC Voltage with the Oscilloscope

Objectives

1. Measure AC voltage with the oscilloscope.
2. Convert between peak, RMS, and peak-to-peak values.

Preparation

The project provides practice in using the oscilloscope to measure AC voltage or amplitude. The lab design allows you to determine if your oscilloscope readings are correct. Voltages are read with an AC meter and then compared with the oscilloscope.

A digital voltmeter, or VOM, is used to determine the output of the signal generator in RMS units. The oscilloscope is then used to measure the same voltage. If the RMS values calculated from the oscilloscope are the same as the voltmeter readings, you can be reasonably sure that you are measuring AC voltage correctly.

Figure 26-1 is the equipment setup. Use the base settings indicated in Experiment 24 for the oscilloscope.

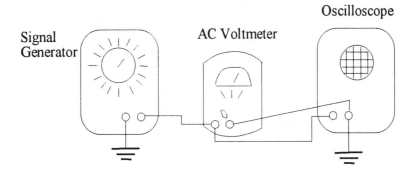

Figure 26-1

Be certain the calibration setting of the vertical sensitivity is at the CALIBRATE position.

Materials

Oscilloscope
Frequency generator
AC voltmeter

Procedure

1. Connect the components as indicated in Figure 26-1. Be sure the lead grounds are connected. This may not be necessary if the internal grounds are common, but it is always good technique to complete the connection externally.

2. Adjust the oscillator frequency to 4 kHz sine wave output. The frequency is not important in this experiment and need not change during the project.

3. Slowly adjust the amplitude control on the oscillator until the meter reads 0.1 V_{RMS}. This is the first voltage in Table 26-1, column 1.

4. Calculate the peak value for an RMS of 0.1 V AC. Enter this in column 2.

5. Calculate the peak-to-peak value and enter in column 3.

6. Adjust the oscilloscope display so that a steady sine wave appears on the screen. The vertical sensitivity (VOLTS/DIV) should be set so the vertical height is between 3 and 8 divisions on the screen. (Check that the calibration control is on CALIBRATE.)

   If the waveform height is less than 3 divisions, its readability will be less precise.

7. Count the divisions from maximum to minimum (peak-peak) and multiply by the VOLTS/DIV setting. Enter this oscilloscope reading in column 4.

8. Divide the peak-to-peak in half to obtain the peak reading and enter this in column 5.

9. Multiply the peak value by 0.707 to obtain the RMS value and enter this in column 6.

10. Is your value in column 6 about the same as the value in column 1? If so, then you have correctly read AC voltage from the oscilloscope.

11. Adjust the oscillator amplitude for the next RMS voltage and complete the data table. Additional spaces are provided if you would like to try some additional voltage readings of your own choice.

Table 26-1 Data for Figure 26-1

Meter Values			Oscilloscope Values		
1	2	3	4	5	6
Adjust Generator	Calculate from RMS	Calculate from Peak	Read Scope	Calculate from Peak-Peak	Calculate from Peak
RMS Volts	Peak	Peak-Peak	Peak-Peak	Peak	RMS
0.1 V					
0.2 V					
0.4 V					
0.7 V					
1.0 V					
1.5 V					
2.0 V					
2.5 V					
Other Optional RMS Volts					

Lab Questions

1. If the VOLTS/DIV CALIBRATE control is not at its CAL position, will the peak reading be too low or too high? _____

2. Does the time or sweep frequency setting affect the vertical voltage reading? _____

3. Prove your answer to question 2. Obtain a sine wave on the screen covering 6 divisions. Move the sweep VAR slightly and notice the vertical amplitude. Did it change? _____ Does the amplitude still cover 6 divisions? _____

    Move the TIME/DIV switch. Does this affect the amplitude? _____

4. What are the sources of measurement error in this experiment?

5. Can an oscilloscope be used to measure current? _____ Justify your answer by explaining how or why.

27 Transformer Fundamentals

Objectives

1. Measure the voltage and current ratio of a transformer.
2. Observe the ability of a transformer to change AC voltage.
3. Determine the current and power requirements of the transformer circuit.

Preparation

The experiment is written for a small 12.6 V power transformer with a single winding in both primary and secondary. The principles can be applied to multiple output and tapped transformers as well.

The experiment looks at the voltage and current ratios, compares primary and secondary power, and identifies sources of measurement error.

The project uses a power transformer at line frequency. It requires only a single primary and single secondary winding, but any step-down power transformer may be used.

Only low, waveform generator voltages will be used. Thus, dangerous line voltages are not exposed on the bench.

Voltage ratio can be used to determine the turns ratio. Resistance measurements will be taken, but cannot be used as an accurate indication of turns. If primary and secondary are not wound with the same size wire (which is often the case), resistance ratio will not determine turns ratio.

Materials

Transformer:
    Low voltage rectifier transformer
        120:12 ratio or similar
        300 mA secondary or larger
Oscillator or waveform generator
One VOM meter
One AC ammeter
One 560 Ω-1/2 watt resistor
One 270 Ω-1/2 watt resistor
One 100 Ω-1 watt resistor

Procedure

Part 1—Voltage and Turns Ratio

1. From the manufacturer's specification, what is the normal input or primary voltage? _____ What is the normal output or secondary voltage? _____

   Calculate the expected voltage ratio based on manufacturer's specification. _____

   $$\text{Voltage ratio} = \frac{\text{Specified Primary Voltage}}{\text{Specified Secondary Voltage}}$$

2. Connect the AC generator to the primary. Set the frequency to 100 Hz and the primary voltage to 10 V$_{RMS}$.

   Measure the secondary voltage. _____

3. From your data in step 2, determine the actual voltage ratio: _____

   $$\text{Voltage Ratio} = \frac{\text{Measured Primary Voltage}}{\text{Measured Secondary Voltage}}$$

4. Compare your measured voltage ratio with the calculation in step 1. _____

5. Increase the primary voltage to 20 V and measure the new secondary voltage. _____

6. Determine the voltage ratio with 20 V input. _____

7. Did the ratio change with different primary voltage? _____

8. Based on your measured voltages, is the transformer a step-up or step-down? _____

9. Since voltage and turns ratio are the same number, indicate the transformer turns ratio for the transformer. _____

   $$\text{Turns Ratio} = \frac{V_P}{V_S}$$

Part 2—Current and Power Requirements

Before applying voltage, determine the current and power requirements of the secondary circuit, with a 100 Ω load resistor.

1. From the transformer specifications, what is the maximum secondary current? _____

2. In Part 1, what was the secondary voltage with 10 V$_{RMS}$ on the primary? _____

3. Divide this secondary voltage by 100 Ω and find the expected secondary current. _____

4. Does your expected secondary current exceed the manufacturer's rating? _____

5. Calculate the power that will be generated by the secondary voltage across the 100 Ω load resistance. _____

   $$P_S = V_S \times I_S$$

6. Is a 1 W resistor large enough? _____ Is a 1/2 W resistor large enough to handle this expected power? _____

7. If your answer is yes, connect a 100 Ω, 1/2 or 1 W resistor to the secondary.

   Measure the secondary voltage and divide by the 100 Ω to obtain the secondary current flowing through the load resistor.

   Measured secondary current = _____

8. Calculate the expected primary current based on the transformer turns ratio and measured secondary current. _____

   $$\text{Calculated primary current} = \frac{\text{Secondary Current}}{\text{Turns Ratio}} = \underline{\qquad}$$

9. Now measure primary current to see if it agrees with your calculation. Disconnect the power. Connect an AC ammeter in series with the primary circuit. (Start with range selector high enough to prevent damaging meter or fuse.)

   Measured primary current = _____

10. Is your primary current reasonably close to calculation, or is it higher than expected? _____

11. A source of error in this situation is the no load primary current. This is the current that flows through the primary winding with no current in the secondary.

    Remove the secondary load and measure primary current. _____

12. Subtract the no load current from the measured current in step 9. _____ Is this value closer to the calculated value? _____

Part 3—Resistance Measurements

Resistance measurements are one means of troubleshooting a transformer. If either the primary or secondary coil is open, the coil will measure maximum ohms, or "open." A good coil will measure the normal winding resistance.

1. Measure the resistance of the primary coil. _____

2. Measure the resistance of the secondary coil. _____

   Do both windings have normal readings? _____

   The physical separation of the windings allows the transformer to be used as an isolation device. The magnetic flux is the only electrical link between input and output.

   Since the primary and secondary windings are separate and isolated from each other, there should be maximum or open resistance between the input and output leads.

3. Measure the resistance between one lead of the primary and one lead of the secondary. _____ Is the transformer in good condition? _____

Lab Questions

1. Transformer "specs" may have a value for maximum secondary or maximum primary current, but not both.

    Why is a current specification for one winding sufficient?

2. From your data in Part 1, is the voltage ratio dependent on the applied voltage?

3. How can you make your transformer a step-up rather than a step-down?

4. Which winding has the most current—the high voltage or the low voltage winding?

5. For an "ideal" transformer, how does the power generated in the primary compare with the power generated in the secondary?

6. Using your data from Part 2, calculate the primary power.

7. Calculate the secondary power.

8. How close are your two values?

9. What electrical characteristic of the transformer contributes to experimental error?

28 *RC* Time Constants

Objectives

1. Observe the DC charge and discharge voltage on a capacitor.
2. Obtain measurement data of charge and discharge voltage.
3. Plot a graph of capacitor voltage at each time constant.

Preparation

1. Refer to Figure 28-1. This is a diagram of the charging circuit configuration. When the switch is closed, current begins to flow through the charging resistor, and the capacitor voltage builds to its maximum value.

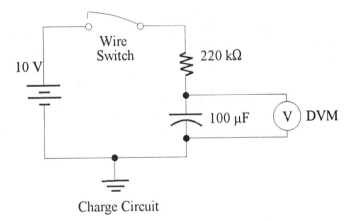

Charge Circuit

Figure 28-1

CAUTION: If the 100 µF capacitor is an electrolytic, it must be connected with the polarity as shown. If it is wired incorrectly, the circuit will not work, and the capacitor can explode.

A DVM with 10 MΩ or higher input is critical to the project. Since the voltmeter resistance is across the capacitor, it will contribute a small measurement error. When the capacitor is fully charged, current will continue to flow through the charging resistor in series with the voltmeter. The actual voltage across the meter/capacitor combination will be somewhat less than the source voltage.

2. Refer to the table for Part 1 (Table 28-1). The first column indicates the Time Constants at which you will measure capacitor voltage. The second column records your calculation of the time in seconds, represented by each Time Constant.

$$\text{(Time in Seconds } = TC \times R \times C)$$

The third column records your measurement data.

Materials

DC power supply
Voltmeter: 10 MΩ input
Resistors:
    One 100 kΩ, one 220 kΩ
Capacitors:
    One 47 μF, 20 WVDC (minimum); one 100 μF, 20 WVDC (minimum)

Procedure

Part 1—Charging Circuit

1. Wire the circuit (Figure 28-1) with voltage supply *off*.

   A wire connector is used as a simple switch to turn the voltage on or off at the circuit.

2. Any voltage on the capacitor should be discharged by connecting the open end of the wire switch to ground.

3. The project must be done with at least two partners. One student is the switch operator and measurement reader. The second student is the timer and monitors a watch or a clock with a seconds indication.

   When the team is ready to begin measurements, the timer announces "start," and the switch operator closes the wire switch. The timer then indicates each Time Constant interval by calling "mark." The measurement reader notes the capacitor voltage and records this value in Table 28-1.

Table 28-1 Data for the Charge Circuit (Figure 28-1)

1	2	3
Time Constant	Calculate Time in Seconds	Measure Capacitor Voltage
0 (Start)		
0.5 TC		
1.0 TC		
1.5 TC		
2.0 TC		
2.5 TC		
3.0 TC		
3.5 TC		
4.0 TC		
4.5 TC		
5.0 TC		
5.5 TC		
6.0 TC		

Part 2—Discharge Circuit

1. Set up the circuit shown in Figure 28-2. With the switch in position A, charge the capacitor to maximum voltage.

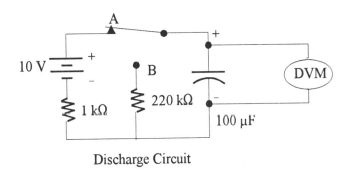

Discharge Circuit

Figure 28-2

2. Move the wire switch to position B and take data for the discharge voltage of the capacitor. The time and method can be the same as Part 1. Voltage measurements are recorded in Table 28-2.

Table 28-2 Data for the Discharge Circuit (Figure 28-2)

1	2	3
Time Constant	Calculate Time in Seconds	Measure Capacitor Voltage
0 (Start)		
0.5 TC		
1.0 TC		
1.5 TC		
2.0 TC		
2.5 TC		
3.0 TC		
3.5 TC		
4.0 TC		
4.5 TC		
5.0 TC		
5.5 TC		
6.0 TC		

Part 3—Plotting a Chart of Charge and Discharge Voltage

1. A labeled graph sheet is included for a plot of DC voltages. Notice that the horizontal axis is labeled in time constants or TC. From your data, add a scale of time in seconds on the horizontal axis.

2. Notice that the vertical axis is labeled in percent of maximum (or applied) volts. Add your own scale for capacitor voltage. The label here will depend on your applied voltage.

   Example: If the applied voltage is 8 V, then the 60% mark would also be labeled 4.8 V (60% of 8 = 4.8).

3. Plot the data from Table 28-1 and Table 28-2. Connect the points smoothly, and you will have a picture of the capacitor voltage in each circuit.

4. The time constant axis and the percent of maximum voltage axis makes your graph a "universal" time constant curve.

   Compare this curve with the graph in your textbook.

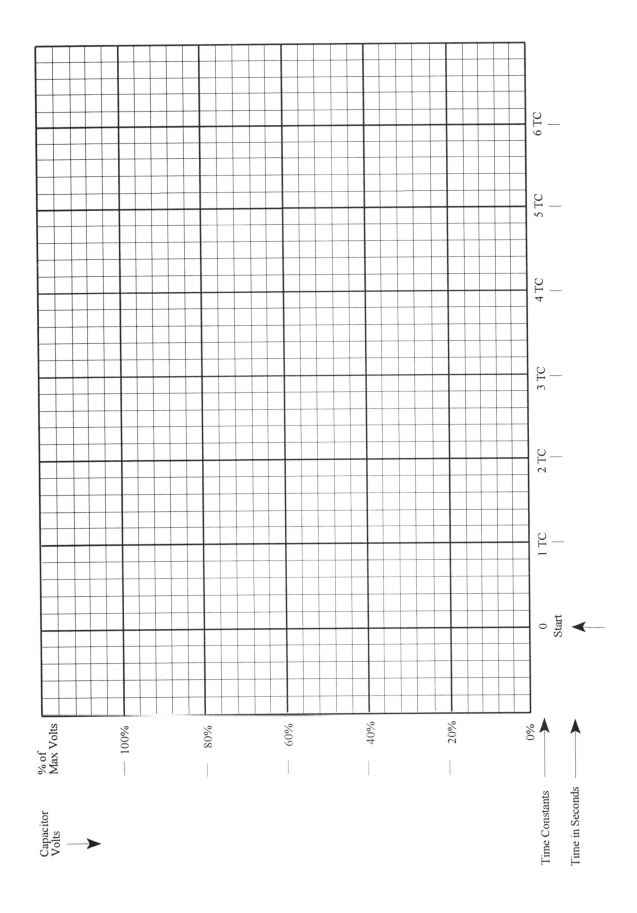

Lab Questions

Note: Your circuit should be set up and ready to operate while you answer these questions.

1. Meter resistance is one source of experimental error. Explain why this problem occurs.

2. Capacitor leakage current can be another problem. How would this create a difference between measured and calculated values. (Remember: Capacitor leakage current must flow through the series resistance.)

3. If the 220 kΩ resistor was replaced with a smaller value, such as 100 kΩ, would the charge time be faster, slower, or no change? _____

4. Replace the 220 kΩ with a 100 kΩ. Operate the circuit and compare the rate of charge with the 220 kΩ. Was your answer correct in question 3? _____

5. If the 100 µF capacitor is replaced with a smaller value, such as 47 µF, would the rate of charge be faster, slower, or no change? _____

6. Return the 220 kΩ to the circuit. Replace the 100 µF capacitor with a 47 µF capacitor. (Be careful of polarity.) Operate the circuit to charge and compare charge time with the original circuit. Is your answer to question 5 correct? _____

7. How does the electrical size of the components affect the charge time?

8. Notice the circuit for obtaining discharge voltage values. The capacitor must first be charged to full voltage through the 1 kΩ resistor with switch in position A.

    How long should you have to wait to fully charge the capacitor in this position? _____

    Assuming the capacitor is fully discharged before connecting the switch to position A, would one second be sufficient time to obtain full voltage? _____

9. Prove your answers to question 8 by calculation and measurement.

29 Capacitance and Capacitive Reactance

Objectives

1. Determine capacitance and capacitive reactance by experiment.
2. Demonstrate the effect of frequency on capacitive reactance.

Preparation

Refer to the circuit arrangements of Figures 29-1 and 29-2. In Figure 29-1, the oscilloscope is used to read the AC voltage across the resistor. In Figure 29-2, the positions of resistor and capacitor have been exchanged, and the scope is reading voltage across the capacitor.

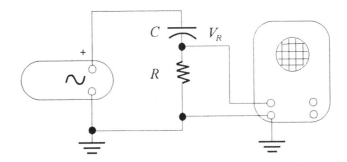

Figure 29-1

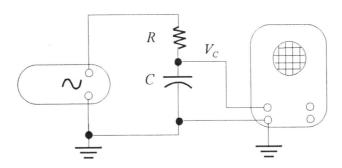

Figure 29-2

Since it is a series circuit, the sequential placement of components does not affect circuit operation. The scope and oscillator can have common grounds in both arrangements.

Ohms is the ratio of voltage to current. In this experiment, you will find the ohms of reactance by measuring V_C and dividing by current.

After finding the capacitive reactance, you can determine the experimental capacitance with the formula:

$$C = \frac{1}{2\pi f X_C} \quad \text{or} \quad C = \frac{.159}{f \times X_C}$$

By comparing three capacitances at three frequencies, you can observe the principles of capacitance in an AC circuit.

Materials

Frequency generator
Oscilloscope
Capacitors:
    One 0.01 μF, one 0.022 μF, one 0.047 μF
One 1 kΩ-1/2 watt resistor

Procedure

1. Build the circuit in Figure 29-1. Set frequency to 1 kHz. Adjust source voltage to 2 V_{p-p}.

2. Read frequency on the oscilloscope and fine tune the generator so the oscilloscope indicates exactly 1 kHz (Experiment 25).

3. Connect the oscilloscope across the current sensing resistor and measure V_R.

    $V_R = $ _____

4. Find the circuit current with Ohm's law. $I = \dfrac{V_R}{R}$ $I = $ _____

5. Exchange the positions of resistor and capacitor (Figure 29-2).

6. Measure the capacitor voltage with the oscilloscope. $V_C = $ _____

7. Divide capacitor voltage by current to obtain the ohms of reactance. $X_C = $ _____

8. Using the experimental value of X_C (from step 7), calculate the experimental capacitance from the formula:

    $$C = \dfrac{1}{2\pi f X_C} \text{ or } C = \dfrac{.159}{f \times X_C}$$

9. Compare your experimental value of capacitance above with the printed or coded value. Are they reasonably close? _____

10. Enter your data for the 0.01 μF capacitor at 1 kHz in Table 29-1.

11. Replace the 0.01 μF capacitor with a 0.022 μF.

    Take the measurements and make the calculations as indicated in Table 29-1.

12. Continue taking measurements and obtaining data to complete Tables 29-2 and 29-3.

Table 29-1 Data for 1 kHz

1	2	3	4	5
Capacitor	Measure I_C	Measure V_C	Find X_C $X_C = \dfrac{V_C}{I}$	Find C $C = \dfrac{.159}{f \times X_C}$
0.01 μF				
0.022 μF				
0.047 μF				

Table 29-2 Data for 2 kHz

1	2	3	4	5
Capacitor	Measure I_C	Measure V_C	Find X_C $X_C = \dfrac{V_C}{I}$	Find C $C = \dfrac{.159}{f \times X_C}$
0.01 μF				
0.022 μF				
0.047 μF				

Table 29-3 Data for 4 kHz

1	2	3	4	5
Capacitor	Measure I_C	Measure V_C	Find X_C $X_C = \dfrac{V_C}{I}$	Find C $C = \dfrac{.159}{f \times X_C}$
0.01 μF				
0.022 μF				
0.047 μF				

Lab Questions

1. From the data of Table 29-1, does the smaller or larger capacitor allow more current to flow at a given frequency? _____

2. Does the smaller or larger capacitor have more ohms reactance at a given frequency? _____

3. Compare each table. Does more current flow at higher or lower frequency? _____

4. From each table, does the value of *capacitance* change with frequency? _____

5. Why is the resistor current the same as the capacitor current?

6. From the arrangement of the circuit in Figure 29-1, what would happen if you placed the oscilloscope directly across the capacitor to measure V_C. This assumes you did not exchange the placement of components and scope ground is not common with generator ground.

    What voltage would appear on the oscilloscope? _____

7. Prove or find the answer to question 6 experimentally. Place the circuit arrangement as in Figure 29-1. Place the oscilloscope directly across the capacitor. With 2 V applied from the generator, what is the oscilloscope reading? _____

    Explain the results.

30 Series and Parallel Capacitance

Objectives

1. Determine experimentally the effect of combining capacitances in series and parallel.
2. Verify the formulas for series and parallel combinations of capacitance.

Preparation

A 1 kΩ resistor is used in series with capacitance to determine the current flowing with an applied AC voltage. The change in current with a change in capacitance will indicate more or fewer ohms of reactance (X_C). This in turn will indicate more or less capacitance (C). The principles of Experiment 29 can then be used to determine whether C is increasing or decreasing when the components are combined in series and in parallel.

A DVM with accurate frequency response up to 500 Hz can be used as the current sensing meter. Alternatively, an oscilloscope can be used as in Experiment 29.

Note: Always check source voltage and frequency after changing any component. It should remain the same throughout the experiment.

Materials

Digital voltmeter
Waveform generator
1 kΩ-1/2 watt resistor
Capacitors:
 One 0.01 μF, one 0.022 μF, one 0.047 μF

Procedure

Part 1—An Experimental Method to Determine the Relative Size of Capacitance

1. Build the circuit of Figure 30-1. Apply 2 V at 500 Hz.

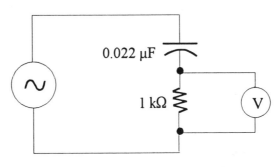

Figure 30-1

2. Measure the voltage across R_S and determine the current flowing in the circuit. ($I = \dfrac{V_R}{R}$)

145

3. Replace the 0.022 μF with a 0.01 μF capacitor. Determine the current flowing with the smaller capacitance. _____

   Was the current with the 0.01 μF more or less than with the 0.022 μF (same applied voltage and frequency)? _____

   Did the smaller capacitor allow more or less current to flow? _____

4. What effect will a larger capacitance have on the circuit current? _____

5. Replace the 0.01 μF with a 0.047 μF capacitor. What happened to the circuit current (increase, decrease, or no change)? _____

Part 2—The Effects of Combining Capacitances in Series or Parallel

Parallel

1. Replace the 0.047 μF capacitor with the 0.022 μF.

2. Once again record the current flowing in the circuit with the 0.022 μF at 500 Hz. _____

3. Place a second 0.022 μF in *parallel* with the first. Notice the current. Did the circuit current increase or decrease when you added a second C in parallel? _____

   From Part 1, does this indicate an increase or decrease in total capacitance? _____

   The result demonstrated here is that adding capacitances in parallel (increases, decreases) _____ the total capacitance of the combination.

4. Place a *third* capacitance in parallel with the first two.

5. Did the current once again increase? _____ Did adding the third capacitance increase the combined capacitance further? _____

Series

6. Remove two of the parallel capacitances, so that you have only the single 0.022 μF in the circuit. Once again check the current with this single capacitance. _____

7. Now add a second 0.022 μF in *series* with the first. (Pay special attention to the PC board or whichever connection system you are using to be sure that it is, in fact, in series.) Measure the current with the additional series capacitance. _____

8. Did the current increase or decrease with the addition of the *series* capacitance? _____

9. You should see a decrease of current with the addition of a series capacitance. From the lessons of Part 1, does this mean a larger or smaller combined capacitance? _____

10. You have demonstrated that combining capacitors in parallel _____ the total capacitance.

    Combining capacitances in series _____ the total capacitance.

Part 3—Experimentally Verify the Calculation of Series and Parallel Capacitance Combinations

Parallel Connection

1. Place two 0.022 µF capacitors in parallel. Use the circuit of Figure 30–1, adding a second capacitor.

2. Measure the current flowing through the combination. _____

3. Measure the voltage across the parallel combination. _____

4. Divide the V_C by I to obtain X_C for the combination. _____

Step 5 is the experimental determination of C_T.

Step 6 is the theoretical determination of C_T.

5. Find C total for the combination by the formula $C_T = \dfrac{.159}{f \times X_C}$: _____

6. Calculate C total for the parallel combination by using the formula $C_T = C_1 + C_2$: _____

   Do the results of steps 5 and 6 compare favorably? _____

Series Connection

7. Connect the two 0.022 µF capacitors in series with the resistance.

8. Measure the current flowing. _____

9. Measure voltage across the *series* combination of capacitance. _____

10. Find the total X_C for the combination. $(X_C = \dfrac{V_C}{I})$ _____

Step 11 is the experimental determination of C_T.

Step 12 is the theoretical determination of C_T.

11. Calculate C_T from the experimental value of X_C: $(C_T = \dfrac{.159}{f \times X_C})$ $C_T = $ _____

12. Calculate C_T from the coded values using $C_T = \dfrac{C_1 \times C_2}{C_1 + C_2}$: $C_T = $ _____

13. How do your experimental and theoretical calculations for C_T compare?

Lab Questions

For the questions below, assume that frequency and applied voltage across the capacitor remains constant. Select *increase*, *decrease*, or *not change* for response.

1. An increase in AC current through the capacitor indicates the capacitance may have _____ .

2. If the current through the capacitor decreases, the reactance must have _____ .

3. If the capacitance becomes smaller, the current will _____ .

4. An increase in current indicates an _____ in X_C.

5. A capacitance inadvertently placed in parallel with the original capacitor will _____ the total C.

31 Series *RC* Circuits

Objectives

1. Analyze a series *RC* circuit by measurement and application of circuit theory.
2. Determine the effect of varying component values and frequency on circuit operation.

Preparation

The experiment analyzes the circuit of Figure 31-1. An electronic or digital voltmeter with an AC frequency response to 2 kHz can be used for direct measurement.

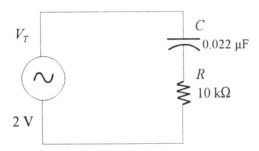

Figure 31-1

Another option is to use the oscilloscope to set the source and measure circuit voltages. All values can be measured in peak-to-peak values for ease of reading. It is not necessary to convert to RMS. The calculations can also be made with peak-to-peak values.

Note: Any calculation that is based on measured values is still called a "measurement" since the result is based entirely on meter readings.

Part 1 is the calculation for the experiment.

Part 2 is the measurement procedure with voltmeter or oscilloscope.

If the DVM is used for the experiment, phase angle can be found from the equation:

$$\theta = \arctan\frac{X_C}{R}$$

If the oscilloscope is used, phase angle can be measured using the procedures of Experiment 32.

Part 3 builds analytical and troubleshooting skills by substituting different components and observing the effect on the circuit.

If an oscilloscope is used for measurements, the placement of resistor and capacitor must be exchanged to keep the grounds common. Figures 31-2 and 31-3 indicate the component placement for each measurement.

Before making any readings, check the source voltage. Be sure it has not changed, and adjust V_T if necessary. This will help prevent errors due to signal generator loading effects.

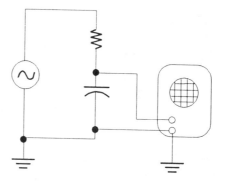

Figure 31-2 Oscilloscope Measurement—Measuring V_C

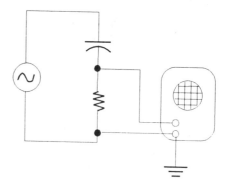

Figure 31-3 Oscilloscope Measurement—Measuring V_R

Materials

Oscilloscope or digital voltmeter
10 kΩ-1/2 W resistor
Capacitors:
    One 0.047 μF, one 0.022 μF, one 0.01 μF

Procedure

Part 1—Calculations

The circuit is Figure 31-1 with the following values.

$R = 10$ kΩ, $C = 0.022$ μF, $V_T = 2$ V$_{p-p}$, $f = 720$ Hz

Calculate the following quantities and enter the data in Table 31-1.

Calculate:

    X_C at the circuit frequency

$$X_C = \frac{1}{2\pi f C}$$

    AC impedance (Z)

$$Z = \sqrt{R^2 + X_C^2}$$

Circuit current (I)

$$I = \frac{V_T}{Z}$$

Resistor voltage (V_R)

$$V_R = I \times R$$

Capacitor voltage (V_C)

$$V_C = I \times X_C$$

Phase angle (θ)

$$\theta = \arctan\frac{-X_C}{R} \text{ or } \arctan\frac{-V_C}{V_R}$$

Part 2—Measurements

1. Set frequency to 720 Hz.

2. Adjust source voltage to 2 V.

3. Measure V_R. _____

4. Find current (measured $\frac{V_R}{R}$). _____

5. Measure V_C. _____

6. Find impedance ($Z = \frac{\text{measured } V_T}{\text{measured } I}$). _____

7. Find $X_C = (\frac{\text{measured } V_C}{\text{measured } I})$. _____

8. Enter measurements in data Table 31-1. Compare your calculated and measured values. Do they fall within acceptable error? _____

Table 31-1 Data for Figure 31-1

Quantity	Calculate	Measure
X_C		
V_C		
V_R		
I		
Z		
θ		

Part 3—Changing Values

Based on your study of AC circuits, predict what will happen to the following values if frequency increases.

1. What will happen to the following:

    V_C _____ V_R _____ I _____ Z _____ θ _____

2. Confirm your answers by completing Table 31-2.

    Measure the values indicated at each of the frequencies listed. Remember, I, Z, and θ are derived from measured values.

    Table 31-2 Data for Circuit with: $C = 0.022\ \mu F$, $R = 10\ k\Omega$, and $V_T = 5\ V$

Quantity	350 Hz	720 Hz	1400 Hz
V_C			
V_R			
I			
Z			
θ			

3. What effect will an increase in capacity have on the following values?

    V_C _____ V_R _____ I _____ Z _____ θ _____

4. Keep frequency constant at 720 Hz, and change C as indicated.

    Obtain the data for Table 31-3 and compare results.

1. **Table 31-3** Data for Circuit with: $f = 720\ Hz$, $R = 10\ kHz$, and $V_T = 5\ V$

Quantity	0.01 μF	0.022 μF	0.047 μF
V_C			
V_R			
I			
Z			
θ			

Lab Questions

Voltage and Impedance

1. You are asked to be sure that the source voltage has not changed when you change a component value. If the circuit current increases, you may find the output of the generator has become lower.

    Why do some AC generators react this way? (Note: If your signal source has a regulated output, you may not notice a change.)

2. Add the voltages across R and across XC. Is the direct addition of V_R and V_C equal to the source voltage? _____

3. Calculate the source voltage using the formula: $V_T = \sqrt{V_R^2 + V_C^2}$

    Does this vector addition give an answer much closer to V_T than question 2? _____

4. Why did the circuit current become larger when you increased the value of capacitance?

5. If you directly add the value of R and X_C, does this equal the total circuit impedance? _____

    Calculate the value of Z by vector addition, using the equation: $Z = \sqrt{R^2 + X_C^2}$

    Is this reasonably close to the derived value found by measurement? _____

6. Why did the circuit impedance increase when you replaced the capacitance with a smaller value?

7. What component caused the circuit current to increase when you raised the frequency from 350 Hz to 720 Hz? _____

    Why did this happen?

Frequency Response

At low frequency

8. At the lower frequency (350 Hz), which voltage is larger, V_C or V_R? _____

9. At low frequency, which quantity is larger, X_C or R? _____

10. According to your answers above, is the circuit phase angle closer to 90° or zero degrees?
_____

11. Does your derived measurement for θ at 350 Hz agree with your answer to question 10?
_____

At mid frequency

12. At a certain frequency, X_C will equal R. The frequency can be calculated from the formula: $f = \dfrac{1}{2 \times \pi \times R \times C}$
    Calculate the frequency for this circuit, where $R = X_C$. _____

13. Is the frequency you calculated approximately the same value as used in the lab project for the middle frequency? _____

14. At this mid frequency, is X_C approximately the same value as R according to your measurements (within about 10 percent)? _____

15. If R and X_C are about the same quantity, what is the phase angle? _____ Does your derived data for θ at 720 Hz confirm this? _____

At high frequency

16. At 1400 Hz which voltage is larger, V_C or V_R? _____

17. According to your answer to question 16, is the phase angle closer to zero or 90°? _____

18. Does your derived measurement for θ at 1400 Hz agree with your answer to question 17?
_____

19. Sketch and label the phasor diagram for V_C, V_R, and V_T at each of the three frequencies. Make the length of the vectors proportional to the voltages at each frequency.

       350 Hz   **720 Hz**   **1400 Hz**

32 Phase Measurements

Objectives

1. Observe the phase difference between current and voltage, displayed on a dual sweep oscilloscope, for resistive and capacitive circuits.
2. Measure the circuit phase angle from the oscilloscope display.
3. Calculate the current and voltage phase angle for the RC circuit and compare with measured values.

Materials

Dual trace oscilloscope
Signal generator
One 0.01 μF capacitor
Resistors: 1/2 watt
    One 5.6 kΩ, one 4.7 kΩ, 1 kΩ

Procedure

Part 1—Resistance Circuit

Preparation

1. Refer to Figure 32-1. This is a resistive circuit, so voltage and current are in phase. You can observe this using a dual trace oscilloscope. Different brands of oscilloscopes can have different names for each of the two "channel" inputs. They may be marked Channel A or CH1 or Ch X or some other indication.) In this lab, they will be referred to simply as A and B to correspond to the location on the measured circuit.

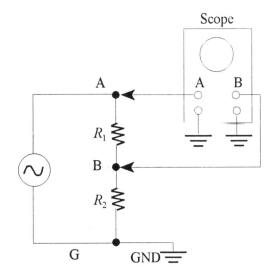

$R_1 = 5.6$ kΩ
$R_2 = 4.7$ kΩ
$f = 1$ kHz
$V_T = 6$ V$_{p\text{-}p}$

Figure 32-1

155

2. Channel A is measuring the total circuit voltage.

Channel B is measuring the voltage across R_2. This resistor can be considered a current sensor for the circuit since any current flowing through it will be represented in magnitude and time by the voltage across the resistor.

The phase angle is zero, so the position of the voltage wave is the same as the current wave. The scope is sensing voltage, but for our purposes, this is a current indicator! Let's see how this is presented on the scope.

Procedure

1. Set up the circuit as shown in Figure 32–1.

2. Select *Channel A input*.

3. Set generator frequency to 1 kHz.

4. Set the oscilloscope sensitivity to 0.1 V/div. Adjust the amplitude at the generator to read 0.6 V_{p-p} on the scope.

5. Set the TIME/DIV to obtain two or three complete waveforms.

6. Place the Channel B sensitivity to 0.1 V/div. (Do not change the channel selector switch. The scope should still be "looking" at input A.)

7. Ground both channels by placing *both* AC-GND-DC switches on GND.

8. Place selector switch on dual channel or "both." You should now have two horizontal lines on the screen. If not, move the vertical position for one or both channels.

9. Move each line so that it is in the center of the screen. This will superimpose both zero reference lines so that they appear to be a single horizontal line.

10. Move the Channel selector switch to A and its mode switch to AC. This presents the input circuit voltage.

11. Move the channel selector to B input and its mode switch to AC. This reads V_{R2} and "represents" the current waveform.

12. Move the selector switch to *both*.

13. You should now see two sine waves superimposed in time, with their zero value centered on the graticule.

14. Notice the phase difference (if any) between V_T and I_T. The phase difference is best measured at the base line. The spacing between the crossing of the two wave forms represents the time or phase difference. Since both of these waveforms should be crossing at the same point, the difference is zero!

Part 2—Capacitance Circuit

Preparation

1. Refer to the *RC* circuit in Figure 32-2.

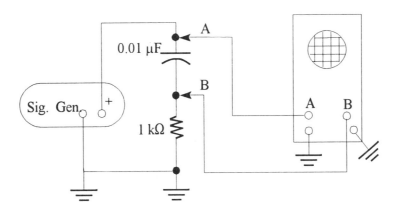

Figure 32-2

Channel A is reading the source voltage or "total" circuit voltage.

Channel B is connected across the current sensing resistor. The two patterns that will be displayed are circuit voltage and a pattern that *represents* the current waveform.

2. Refer to the data table for this part of the experiment (Table 32-1).

   Column 1 lists a set of frequencies that will be used for observation of phase difference.

   Column 2 is the calculation of X_C for each frequency.

   Column 3 is your calculation of phase angle for each frequency.

   Column 4 is your measurement of the phase angle from the lab circuit. (This should be compared with your calculated values).

3. Calculate X_C for each frequency using the component values in your circuit, and enter the capacitive reactance (X_C) in column 2.

4. Calculate the phase angle for these frequencies and enter in column 3.

5. Measure the phase difference using the procedure below.

Procedure

1. Refer to Channel A input and set its sensitivity to 0.1 VOLTS/DIV.

2. Wire the circuit. Set frequency to 1 kHz and the generator voltage to 0.6 V_{p-p} as measured on the oscilloscope.

3. Select Dual Channel Input.

   Adjust Channel B VOLTS/DIV and its variable control to whatever is required to make waveform B the same height as waveform A on the screen. Since you are not measuring voltage here, this *un*calibration is not important. It may, however, make observing the phase angle easier.

4. Adjust the *time* vernier or variable control so that the sine wave pattern on the screen covers exactly 4 cm horizontally. You may need to combine this setting with the TIME/DIV control.

   Pattern should look similar to Figure 32-3.

   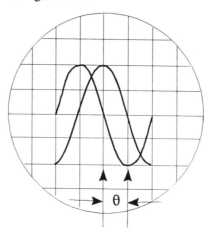

   Figure 32-3 Example Measurement (approx. 90°)

5. Note: One full waveform covers 4 time divisions. So each division contains 360/4 or 90°. (If the sine wave covered 8 divisions, then each major division on the scope would represent 45°.)

   How many *minor* divisions are contained in each major division? Count them! _____
   Assume your scope contains 5 minor divisions for each large division. Then each minor division is "worth" 90/5 or **18°**.

6. Measure the phase angle difference for your voltage and current pattern. Be sure that both waveforms are centered on the screen. Notice where each wave crosses the horizontal time line. Count the number of minor divisions between each wave (include any fractions of a division). Multiply this by 18°. _____

   Enter this angle in Column 4 of the data table.

7. Calculate the phase angle by using the formula: $A = \arctan \dfrac{X_C}{R}$. Enter this value in the calculation cell for 1 kHz. Compare this with your measured value.

8. Complete all measurements and calculations and enter into the data table.

Table 32-1 Data Table for Figure 32-2

1	2	3	4
Frequency	Calculate X_C	Calculate Phase Angle	Measure Phase Angle
1 kHz			
2 kHz			
5 kHz			
20 kHz			
50 kHz			

Lab Questions

Resistance Circuit

1. If you change the input frequency, would you expect the phase difference between the two waveforms to change? _____

2. If the frequency changed, would you expect the width of the waveform pattern on the screen to change? _____

3. Change the input frequency to 2 kHz. Did the phase angle change? _____ Do both waveforms still cross the zero axis at the same point? _____

4. From your observations, does frequency affect the phase angle in a pure resistive circuit? _____

5. If you change the voltage input, would you expect the phase angle to change? _____

6. Slowly increase and decrease the generator output voltage, while observing the pattern. Did the phase difference change or did it remain the same? _____

Capacitive Circuit

1. If you change the input frequency, would you expect the phase difference between the two waveforms to change? _____

2. Change the input frequency to 2 kHz. Did the phase angle change? _____ Do both waveforms still cross the zero axis with the same separation? _____

3. From your observations on the screen, does frequency affect the phase angle in a capacitive circuit? _____

4. If you change the voltage input, would you expect the phase angle to change? _____

5. Slowly increase and decrease the signal generator voltage while observing the pattern. Did the phase difference change or did it remain the same? _____

33 Parallel *RC* Circuits

Objectives

1. Measure AC source and branch currents in a parallel RC circuit.
2. Determine impedance by measurement and application of circuit theory.
3. Observe the effects of changing frequency and component values on circuit operation.

Preparation

The experiment uses a current sensing resistor for measuring branch and source currents. A single resistor is placed in series with the signal generator. Refer to Figure 33-1. Voltage is measured across this resistor, and by Ohm's law you can determine the current.

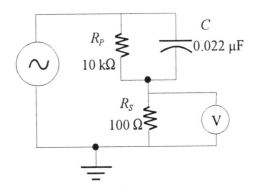

Figure 33-1

The measurement can be made with an oscilloscope, or digital voltmeter, with high frequency response.

To measure capacitor current, remove one of the parallel resistor leads from the breadboard. No current will flow through that resistor, and the current indication will represent capacitor current only.

To measure the resistor current, replace the resistor lead and lift one capacitor lead from the breadboard. No current will flow through the capacitor, and current indication will be for the parallel resistance only.

To measure the total current, connect both components, and the current indication will represent the combined currents, or I_T.

The value of R_S is small enough that it can be ignored in making circuit calculations or measurements.

Before taking any readings, adjust the source voltage, if necessary, to keep it constant. Any change in total current may cause the generator to change its voltage output.

Materials

Signal generator
Oscilloscope
Resistors: 1/2 watt
    One 10 kΩ, one 4.7 kΩ
Capacitors:
    One 0.01 µF, one 0.022 µF

Procedure

Part 1—Calculations

Begin the project by making the following calculations for the circuit of Figure 33-2. Use the following circuit values.

$V_T = 2$ V, $f = 350$ Hz, $R = 10$ kΩ, and $C = 0.022$ µF

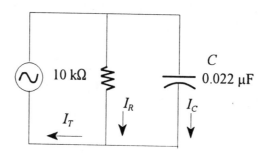

Figure 33-2

1. Capacitive reactance at 350 Hz. $X_C = $ _____

2. Resistor current $I_R = $ _____

3. Capacitor current $I_C = $ _____

4. Source Current $I_T = $ _____

5. Circuit Impedance $(Z = \dfrac{V_T}{I_T})$ $Z = $ _____

6. Phase angle $\theta = $ _____

Enter your circuit calculations in Table 33-1, column 2.

Part 2—Measurements

If you are using a digital voltmeter, set the source to 2 V_{RMS} and read the voltage values in RMS units.

You can directly measure the voltage across the parallel branch as well as voltage across the current sensing resistor.

If you are using an oscilloscope, set the AC supply to 2 V_{p-p}. All circuit readings may be in peak-peak values as well. Current is derived by measuring voltage across the series resistance and dividing by 100 Ω. The source voltage may be used as parallel branch voltage since the series I_R loss is small.

Set up the circuit as shown in Figure 33-1. Use the measurement techniques discussed in the preparation section.

Measure:

1. Parallel branch voltage $\quad\quad\quad\quad\quad\quad V_P = $ _____

2. Parallel resistor current $\quad\quad\quad\quad\quad\quad I_R = $ _____

3. Capacitor current $\quad\quad\quad\quad\quad\quad\quad\quad I_C = $ _____

4. Capacitive reactance ($X_C = \dfrac{V_P}{I_C}$) $\quad\quad X_C = $ _____

5. Total branch currents $\quad\quad\quad\quad\quad\quad I_T = $ _____

6. Circuit Impedance ($\dfrac{V_T}{I_T}$) $\quad\quad\quad\quad Z = $ _____

7. Phase angle ($\theta = \arctan\dfrac{I_C}{I_R}$) $\quad\quad \theta = $ _____

8. Enter data from steps 1 through 7 in the Table 33-1, column 3.

9. Increase the frequency to 730 Hz. Check source voltage. Repeat the measurements and enter in column 4.

10. Increase the frequency to 1400 Hz. Check source voltage. Take measurements and complete data for column 5.

Table 33-1 Parallel RC Circuit

1	2	3	4	5
Freq.	Calculate 350 Hz	Measure		
		350 Hz	730 Hz	1400 Hz
V_T	2 V	2 V	2 V	2 V
X_C				
V_P				
I_R				
I_C				
I_T				
Z				
*θ				
**Compare				

* Phase angle derived from $\arctan \dfrac{I_C}{I_R}$

**Do calculated and measured values compare at 350 Hz? _____

Part 3—Troubleshooting

Adjust generator frequency to 730 Hz.

1. If the value of parallel resistance changed to a lower value, what do you think would happen to the total circuit current? _____

   What would happen to circuit impedance? _____

2. Change the value of R_P to a resistance about half as large. (You can make the selection. Pick any resistor with about half the resistance as the original.)

3. Did the circuit *current* increase or decrease when you lowered R_P? _____

   Based on this result, did the circuit *impedance* increase or decrease when you lowered R_P? _____

4. If the value of capacitance decreased, what would happen to total circuit current? _____

   What would happen to circuit impedance. _____

5. Replace the original resistor.

   Remove the capacitor and select a new component with about half the capacitance value as the original (0.01 µF). Change C to this new value.

   Check source voltage and frequency.

   Read the new source current. _____

   Based on the new current, determine the new Z. _____

   Were your predictions for question 4 correct? _____

6. If one of the parallel branch components became open, what would happen to circuit current? _____

   What would happen to the circuit impedance?

7. Remove one lead of the capacitor and measure I_T. _____

   Calculate the circuit impedance based on this new current. _____

   Was your prediction correct? _____

Lab Questions

1. Could a branch current be obtained by placing an AC ammeter in the branch in series with the component? _____

    Would there be an error introduced by this placement, and what would cause it?

2. When you made the capacitor *smaller*, the circuit impedance became *larger*. Why did this happen?

3. When you made the parallel *resistance smaller*, the circuit *impedance* became *lower*. (The opposite result from making the capacitance smaller.) Why?

4. What happened to the circuit impedance when you opened the capacitor branch?

    What caused this result?

5. Draw the schematic for a circuit that would allow you to obtain measured phase angle with the dual sweep oscilloscope.

34 Inductance and Inductive Reactance

Objectives

1. Experimentally determine the inductance and inductive reactance of a coil.
2. Plot a response curve of inductive reactance over a range of frequencies.
3. Determine the effect of combining inductances in series and in parallel.

Preparation

Refer to the circuit of Figure 34-1. The inductance is a 22 mH coil with an internal resistance between 50 Ω and 150 Ω. This resistance is low enough, compared to X_L over the range of frequencies, that it can be ignored for this experiment.

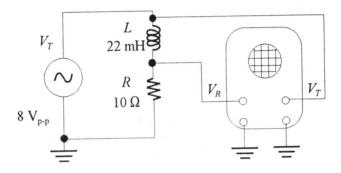

Figure 34-1

The series resistor is the current sensing component.

A dual trace scope is used to identify source voltage on Channel B and resistor voltage on Channel A. A single trace scope can be used if the source voltage is checked after each change in circuit frequency.

In Part 1, you will measure V_L and I_L at several frequencies, calculate X_L, and observe the change in X_L with frequency.

In Part 2, you will plot a graph of X_L versus frequency.

In Part 3, you will measure the effect of combining inductances in series and parallel.

Materials

Frequency generator
Single or dual trace oscilloscope
Two 22 mH inductors
One 10 Ω-1/2 watt resistor

Procedure

Part 1—Inductive Reactance

1. Build the circuit of Figure 34-1.

2. Set the frequency to 40 kHz and source voltage to 8 V_{p-p}.

3. Measure voltage drop across the resistor. $V_R = $ _____

4. Determine circuit current. ($I = \dfrac{V_R}{R}$) $I = $ _____

5. Exchange the positions of resistor and coil.

6. Measure voltage drop across the inductor. $V_L = $ _____

7. Calculate ohms of inductive reactance. ($X_L = \dfrac{V_L}{I}$) $X_L = $ _____

8. Complete the data table (Table 34-1) as indicated. Adjust frequency to each value and maintain source voltage constant.

Table 34-1 Data for Figure 34-1

1	2	3	4
Frequency	Measure V_R	Measure Current ($\dfrac{V_R}{R}$)	Measure X_L ($\dfrac{V_L}{I}$)
10 kHz			
20 kHz			
30 kHz			
40 kHz			
50 kHz			
60 kHz			
70 kHz			
80 kHz			

Part 2—Plotting a Graph of X_L Versus Frequency

Label the graph with frequency on the horizontal axis and inductive reactance on the vertical axis. Plot the data from columns 1 and 4 (Table 34-1) on your graph.

You can begin the X axis at zero frequency, but you should not plot a point there since you did not actually have any data for zero frequency. After completing the curve, you may *extrapolate* with a dashed line to indicate where the curve may have been if you had taken data.

Part 3—Series and Parallel Combinations of Inductance

1. From your data at 40 kHz, calculate the inductance of your coil from the equation:

$$L = \frac{X_L}{2\pi f}$$

2. How close is your experimental inductance to rated value? _____

   (Commercial coils typically have tolerances from 5% to 20%.)

Parallel Inductance

3. Place a second 22 mH coil in parallel with the first (Figure 34-2).

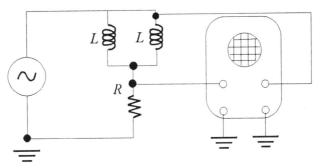

Figure 34-2

4. Set frequency at 40 kHz and measure current.

5. Find X_L of the combined inductance from your measured current and V_T. _____

$$X_L = \frac{V_T}{I}$$

6. Calculate experimental inductance from X_L. _____

$$L = \frac{X_L}{2\pi f}$$

7. Calculate theoretical inductance of two parallel 22 mH coils.

Series Inductance

8. Build a series combination of two coils and the series resistor (Figure 34-3).

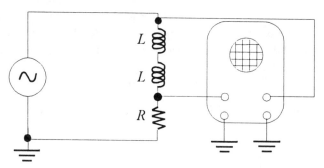

Figure 34-3

9. Find the total inductance of the series combination by using the method of steps 4-6. _____

10. Calculate theoretical inductance of the two series 22 mH coils. _____

11. Compare steps 9 and 10. _____

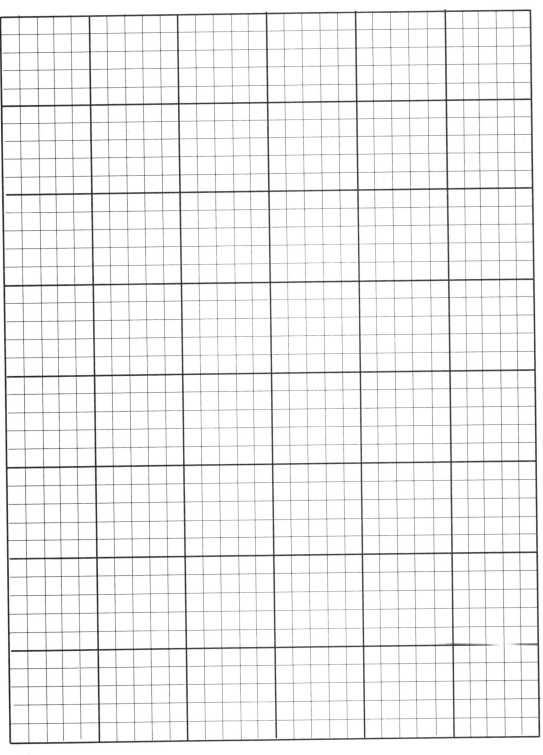

Lab Questions

1. From your data in Table 34–1, did you find more current or less current as the frequency increased? _____ What caused this change in current?

2. Did *inductance* change as frequency changed? _____ It should have remained reasonably constant. Why?

3. Did *inductive reactance* change as frequency changed? _____ What caused this change in X_L with frequency?

4. Did you need to adjust the source voltage during the progress of the experiment to maintain V_T constant? _____ If so, what caused this?

   What was the reason for the generator voltage wanting to change?

5. When the frequency doubled from 20 kHz to 40 kHz, did X_L double as well (approximately)?

6. A certain coil has an $X_L = 100\ \Omega$ at 5 kHz. How many ohms of reactance will it have at 10 kHz?

35 Series *RL* Circuits

Objectives

1. Measure current, voltage, impedance, and phase angle of a series *RL* circuit.
2. Determine the effect of frequency on circuit values.

Preparation

Part 1 is the calculation of circuit values at one frequency. It allows you to compare circuit theory with experimental results.

Parts 2 and 3 measure circuit values over a range of frequencies. You will be able to observe the changes in values as the circuit characteristic changes from resistive to inductive.

Some values are calculated in two ways as a check on calculation and an illustration of theory.

Materials

Frequency generator
Dual trace oscilloscope
One 33 mH inductor
One 10 kΩ-1/2 watt resistor

Procedure

Part 1—Calculations

Refer to the circuit of Figure 35–1.

$R = 10$ kΩ, $L = 33$ mH, $f = 50$ kHz, and $V_T = 4$ V$_{\text{p-p}}$

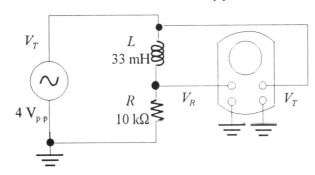

Figure 35–1

Calculate the following quantities for the circuit. Enter your calculations for steps 1 through 6 in Table 35-1.

1. Inductive reactance for the 33 mH coil at 50 kHz. $X_L = $ _____

   $X_L = 2\pi f L$

2. Circuit impedance $Z = $ _____

   $Z = \sqrt{X_L^2 + R^2}$

3. Circuit current $I = $ _____

   $I = \dfrac{V_T}{Z}$

4. Inductor voltage $V_L = $ _____

   $V_L = I \times X_L$

5. Resistor voltage $V_R = $ _____

   $V_R = I \times R$

6. Circuit phase angle $\theta = $ _____

   $\theta = \arctan\dfrac{V_L}{V_R}$ or $\arctan\dfrac{X_L}{R}$

7. Confirm calculation of component voltages. $V_T = $ _____

   $V_T = \sqrt{V_L^2 + V_R^2}$

8. Confirm calculation of circuit impedance. $Z = $ _____

   $Z = \dfrac{V_T}{I}$

Part 2—Measurements

1. Obtain the coil for the project and measure its DC resistance. Record its value.

    $R_{coil} = $ _____

2. Build the circuit of Figure 35–1.

    Set the generator frequency to 50 kHz and the source voltage to 4 V_{p-p}.

    Measure the quantities indicated below and enter your measured values in Table 35–1.

3. Confirm the frequency setting of the generator by measuring frequency on the oscilloscope.

    $f = $ _____

4. Measure V_R with the oscilloscope. $V_R = $ _____

5. Determine circuit current, I. $I = $ _____

6. Determine impedance, Z. $Z = $ _____

    $$Z = \frac{V_T}{I}$$

7. Measure phase angle with the oscilloscope (Experiment 32). $\theta = $ _____

8. Exchange the positions of R and L.

9. With the coil at ground, measure V_L. $V_L = $ _____

10. Determine phase angle from the formula. $\theta = $ _____

    $$\theta = \arctan \frac{V_L}{V_R}$$

11. Confirm V_T.

    $$V_T = \sqrt{V_L^2 + R^2}$$

Table 35-1 Data to accompany Figure 35-1

Quantity	Calculate	Measure	Compare
Frequency	/////		/////
X_L		/////	/////
Z			
I			
V_L			
V_R			
V_T			
θ			

Part 3—Frequency Response

1. Configure the circuit with resistance at ground as in Figure 35-1.

2. Adjust the frequency to 10 kHz. Measure V_R.

3. Measure V_R at each of the frequencies indicated in Table 35-2.

   Adjust source voltage if necessary to keep V_T constant.

4. Reverse the position of the coil and resistor. This will keep generator and oscilloscope grounds common for the next procedure.

5. Measure V_L at each frequency.

6. Calculate phase angles using $\arctan\dfrac{V_L}{V_R}$.

7. Complete the data table for the range of frequencies indicated.

Table 35-2 Frequency Response Data for Figure 35-1

Frequency	Measure V_R	Measure V_L	Calculate $\arctan\dfrac{V_L}{V_R}$
10 kHz			
30 kHz			
50 kHz			
75 kHz			
100 kHz			
200 kHz			

Lab Questions

1. What was the reactance of the coil at 50 kHz? _____

2. Was X_L about the same value as the series resistance? _____

3. If X_L and R in series are the same value, what is the circuit phase angle? _____

4. Did the phase angle measure about 45° with the oscilloscope? _____

5. Did the calculation indicate about 45° using the formula $\arctan\dfrac{V_L}{V_R}$? _____

6. Calculate the coil's quality factor, Q, at 50 kHz: coil $Q = \dfrac{X_L}{R_{coil}}$ _____

7. Calculate Q for the circuit: circuit $Q = \dfrac{X_L}{R_{coil} + R_S}$ _____

8. Does the coil Q stay the same throughout the experiment or does it change with frequency? _____ Why?

9. Why does V_L change as the frequency changes?

10. If R stays constant throughout the experiment, why does the value of V_R become smaller as the frequency increases?

11. Why is V_L much smaller than V_R at low frequencies?

12. Why is V_L much larger than V_R at high frequencies?

13. Draw the voltage vector diagram for the circuit at 10 kHz, 75 kHz, and 200 kHz. Draw the diagrams approximately to scale and indicate phase angle.

Voltage Vector Diagrams for RL Circuits

10 kHz   75 kHz   200 kHz

36 Parallel RL Circuits

Objectives

1. Measure current and impedance of a parallel RL circuit.
2. Observe the effect of frequency change on parallel RL circuits.

Preparation

The first section is the calculation and measurement of circuit quantities at one frequency. Some values are calculated in two different ways to verify circuit theory.

   Current through each branch will be measured with the same technique used previously with other parallel AC circuits.

   The second section looks at the circuit measurement over a range of frequencies. You can observe the circuit response as the characteristics change from inductive to resistive.

Materials

Resistors: 1/2 watt
   One 100 Ω, one 6.8 kΩ
One 22 mH coil
Signal generator
Single or dual trace oscilloscope

Procedure

Part 1—Calculations

Refer to the circuit in Figure 36–1.

$f = 50$ kHz, $V_T = 4$ V, $L = 22$ mH, $R_P = 6.8$ kΩ

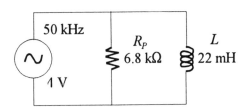

Figure 36–1

Calculate the following values and enter in Table 36–1.

1. $X_L = 2\pi f L$

2. $I_L = \dfrac{V}{X_L}$

179

3. $I_R = \dfrac{V}{R}$

4. $I_T = \sqrt{I_L^2 + I_R^2}$

5. $Z = \dfrac{V}{I_T}$

6. Circuit phase angle, $\theta = \arctan \dfrac{I_L}{I_R}$

Part 2—Measurements at 50 kHz

1. Build the circuit of Figure 36-2. R_S is the current sensing resistor. Since it is in series with the source voltage, any circuit current flowing can be determined by measuring its voltage and dividing by 100.

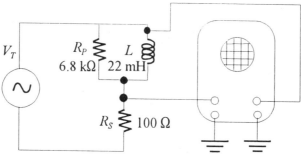

Figure 36-2

2. Apply 4 $V_{p\text{-}p}$ at 50 kHz. Measure I_T, the combined branch currents, and enter in Table 35-1.

    $I_T =$ _____

3. Measure the parallel resistor current by lifting one lead of the coil. Since the coil is no longer in the circuit, the current flowing must be the parallel resistance current.

    $I_R =$ _____

4. Use the same technique to measure the inductor current. Replace the coil lead. Lift one lead of the parallel resistor and measure I_L.

    $I_L =$ _____

5. Confirm your measured value of I_T by determining total circuit current based on the branch currents.

    $I_T = \sqrt{\text{measured } I_R^2 + \text{measured } I_L^2} =$ _____

6. Determine circuit impedance.

    $Z = \dfrac{V_T}{I_T} =$ _____

7. Determine circuit phase angle, θ.

$$\theta = \arctan \frac{I_L}{I_R}$$

Table 36-1 Data for Figure 36-1 and Figure 36-2

Quantity	Calculate	Measure
X_L		
I_L		
I_R		
I_T		
Z		
θ Angle		

Part 3—Measurements Over a Range of Frequencies

1. Connect all components to the board. Reduce the frequency to 10 kHz.

2. At 10 kHz, measure I_T, I_R, I_L, and enter into Table 36-2.

3. From these measurements, determine the value of θ, Z, and I_T based on measured I_R and I_L.

4. Increase the frequency to each value indicated and complete the table.

Table 36-2 Measurements Over a Range of Frequencies

1	2	3	4	5	6	7	8
Frequency	Measure V_T	Measure I_T	Measure I_L	Measure I_R	Calculate Z	Calculate θ	Calculate I_T
10 kHz							
30 kHz							
50 kHz							
75 kHz							
100 kHz							
150 kHz							

Lab Questions

From Table 36-1:

1. Did the measured value of I_T agree with the vector sum of I_R and I_T? _____

2. Would the direct addition of I_R and I_L equal I_T? _____

3. What was the phase angle at 50 kHz? _____

4. Sketch a vector or phasor diagram for the two currents and the indicated angle.

From Table 36-2:

5. At low frequency, which parallel component had the highest current? _____ Why did this occur at low frequency?

6. At high frequency, which parallel component had the highest current? _____ What was the reason?

7. As the frequency became higher, did circuit current increase or decrease? _____ What was the reason for this change?

8. Was the phase angle greater at low frequency or at high frequency? _____ Why?

37 Series *RLC* Circuits

Objectives

1. Measure and calculate the current, voltage, and phase characteristics of a series *RLC* circuit.
2. Observe the effect of frequency changes on circuit operation.

Preparation

The experiment focuses on series *RLC* circuits over a range of frequencies. The frequencies cover the spectrum below, through, and above resonance. The circuit can be observed acting inductively and then capacitively.

The order of placement for components depends on the particular voltage being measured. *L* and *C* should be placed in sequence and not separated by the resistor. This will allow you to take a voltage measurement across both *L* and *C*, with one of these at ground.

Remember, the oscilloscope, source, and circuit grounds must all be common.

It is suggested that you take measurement data by adjusting frequency for each quantity and proceeding horizontally across each row of the data table. You can "fill" the table vertically as an option, but this will require more rearranging of components.

The reactive nature of the circuit is indicated in several ways. If V_L or X_L is larger than V_C or X_C, the circuit is inductive. If the phase angle calculates positive, it also shows that the inductive quantities are larger than the capacitive, and the circuit is inductive. The opposite also applies.

Materials

Dual sweep oscilloscope
Frequency generator
100 mH inductor
1000 Ω-1/2 watt resistor
0.01 μF capacitor

Procedure

Part 1—Calculations

1. Refer to the circuit of Figure 37-1.

    $L = 100$ mH, $C = 0.01$ µF, $R = 1$ kΩ, and $V_T = 2$ V

    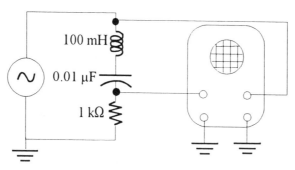

    Figure 37-1 Measure V_R and θ

    The oscilloscope is used to monitor source voltage and measure component voltage. CH1 is the measurement probe. CH2 is the generator probe.

2. Calculate the following quantities at a frequency of 3 kHz. Enter your results in column 2 of Table 37-1.

    Calculate: $X_L, X_C, Z, I, V_C, V_L, V_R$

    $X_L = 2\pi fL$

    $X_C = \dfrac{1}{2\pi fC}$

    $Z = \sqrt{(X_L - X_C)^2 + R^2}$

    $I = \dfrac{V_T}{Z}$

    $V_C = I \times X_C$

    $V_L = I \times X_L$

    $V_R = I \times R$

3. Calculate X_T or the combined reactive ohms. $(X_T = X_L - X_C)$ _____

4. Calculate V_X or the combined reactive voltage. ($V_X = V_L - V_C$) _____

5. Calculate the circuit phase angle including polarity, plus or minus.

$$\theta = \arctan\frac{X_L - X_C}{R}$$ _____

Part 2—Measurements

1. Build the circuit of Figure 37-1. Adjust source voltage to 2 V.

2. Measure V_R at each frequency indicated in Table 37-1 and enter the data.

3. Measure the phase angle with a dual sweep oscilloscope using the same methods as Experiment 32. Notice whether V_R crosses the zero reference line before or after V_T.

    If V_R crosses *before* V_T, current is leading, phase angle is negative, and the circuit is capacitive.

    If V_R crosses *after* V_T, I is lagging, phase is positive, and the circuit is inductive.

    Measure phase angle with sign at each frequency and enter data.

4. Rearrange the circuit as in Figure 37-2. Measure V_L at each frequency.

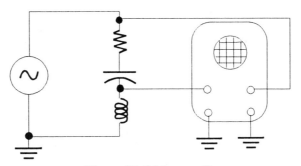

Figure 37-2 Measure V_L

5. Configure the circuit as in Figure 37-3 and measure V_C at each frequency.

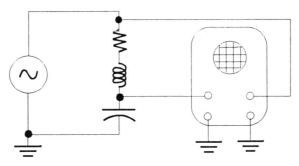

Figure 37-3 Measure V_C

6. Connect components as in Figure 37-4, and measure V_X across inductor and capacitor.

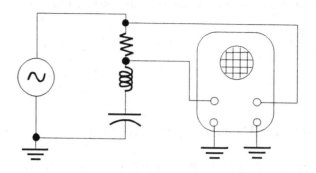

Figure 37-4 Measure $(V_L - V_C)$

7. From the *measurements* above, find the values of: I, X_L, X_C, Z, and X_T. Enter these values.

$$I = \frac{V_R}{R} = \underline{\qquad}$$

$$X_L = \frac{V_L}{I} = \underline{\qquad}$$

$$X_C = \frac{V_C}{I} = \underline{\qquad}$$

$$Z = \frac{V_T}{I} = \underline{\qquad}$$

$$X_T = \frac{V_L - V_C}{I} = \underline{\qquad}$$

8. Calculate the phase angle for each frequency from the measured values of V_L, V_C, and V_R and record values.

$$\theta = \arctan\frac{V_L - V_C}{V_R} = \underline{\qquad}$$

Table 37-1 Series *RLC* Circuit

1	2	3	4	5	6	7
Quantity	Calculate At 3 kHz	Measure at Frequency				
		3 kHz	4 kHz	5 kHz	6 kHz	8 kHz
V_R						
V_L						
V_C						
V_T						
V_X						
I (mA)						
X_L						
X_C						
Z						
X_T						
*θ						
**θ						

*Phase angle derived from $\theta = \arctan \dfrac{V_L - V_C}{V_R}$

**Phase angle measured with oscilloscope

Lab Questions

1. Did your calculated and measured values at 3 kHz frequency compare favorably? _____ Since they were probably not exactly the same, what were the sources of difference?

2. According to your data, what frequency (approximately) created an inductive reactance equal to the capacitive reactance? ($X_L = X_C$) _____

3. What is the name of this special frequency? (Its symbol is f_r.) _____

4. Was the magnitude of the phase angle large or small at f_r? _____

5. What is the sign of the phase angle below f_r? _____ Above f_r? _____

6. At what frequency was current the greatest? _____

7. Explain the reason for your observation in question 6.

8. Consider the data for 6 kHz. What two-element RL circuit would have exactly the same impedance and phase angle as this circuit at 6 kHz? _____ Draw the circuit.

9. When you measured DC circuits with a bench meter, you simply placed the probes of the voltmeter across the leads of any resistor to be measured. What happens if you do that with a grounded oscilloscope?

10. What happened to X_T as the input frequency changed from low through f_r and then to high?

38 Series Resonance

Objectives

1. Observe the operation of a series *RLC* circuit at resonant frequency.
2. Obtain data for the frequency response curve and draw the graph.
3. Analyze the graph to find bandwidth, half-power points, and circuit Q.
4. Observe the effect of component values on the resonant frequency.

Preparation

The lab explores the special characteristics of an *RLC* circuit at the condition when the reactances are equal in magnitude. Since inductor voltage leads the current 90° and capacitor voltage lags the same current by 90°, V_L and V_C will be 180° out of phase. This creates the interesting situation where a voltage measured across both X_L and X_C will measure zero, but the individual voltages will measure significant values, and each of these RMS voltages may be higher than the source!

The first three sections contain the calculations and measurements of circuit parameters.

The fourth section develops the data for a graph of the resonant frequency response, so you can plot a curve of current and impedance versus frequency.

In Part 5, you will determine bandwidth, Q, and half-power points by analyzing the graph.

In Part 6, you will examine the effect of different component values on the resonant frequency.

A dual channel oscilloscope is indicated for all measurements. Channel 1 measures circuit values. Channel 2 is monitoring the source voltage.

Materials

PC assembly board
AC waveform generator
Dual or single trace oscilloscope
Inductors:
    One 100 mH, one 56 mH, one 33 mH
Capacitors:
    One 0.01 μF, one 0.022 μF
Resistors: 1/2 watt
    One 1000 Ω, one 560 Ω, one 270 Ω

Procedure

Part 1—Circuit Calculations

1. Refer to the circuit of Figure 38-1.

   $L = 100$ mH, $C = 0.01$ μF, $R = 560$ Ω, and $V_T = 2$ V$_{p-p}$

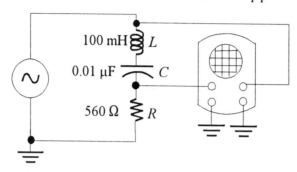

Figure 38-1

Calculate the following quantities for this circuit. If you perform the calculations in the sequence suggested, each solved answer will lead to the next. Enter these values in the second column of Table 38-1.

Resonant frequency $(f_r) = \dfrac{1}{2\pi\sqrt{LC}}$

Inductive reactance $(X_L) = 2\pi f_r L$

Capacitive reactance $(X_C) = \dfrac{1}{2\pi fC}$

Net reactance $(X_T) = (X_L - X_C)$

Circuit impedance $(Z) = \sqrt{X_L^2 + R^2}$

Circuit current $(I) = \dfrac{V_T}{Z}$

Voltage across capacitor $(V_C) = I \times X_C$

Voltage across inductor $(V_L) = I \times X_L$

Net reactive voltage $(V_L - V_C) = I \times X_T$

Voltage across resistance $(V_R) = I \times R$

Phase angle $\theta = \arctan\dfrac{V_L - V_C}{V_R}$

Circuit quality factor $(Q) = \dfrac{X_L}{R_T}$

Bandwidth $(BW) = \dfrac{f_r}{Q}$

Lower half-power frequency $(f_1) = f_r - \dfrac{BW}{2}$

Upper half-power frequency $(f_2) = f_r + \dfrac{BW}{2}$

Part 2—Observation of Resonant and Cutoff Frequencies

1. Build the circuit and set the supply to 2 V_{p-p}. Maintain the supply voltage constant.

   Find the *resonant frequency* (f_r), using the following method.

   Observe the voltage pattern across the resistance (V_R). Adjust the oscillator tuning dial above and below the calculated f_r. Stop at the frequency where the scope pattern is maximum. (You may need to adjust the VOLTS/DIV switch to keep the entire picture on the screen.)

   When you reach the frequency at which V_R is maximum (indicating I is maximum), you have found the resonant frequency.

   Read the oscillator setting and enter the resonant frequency here and in Table 38–1.

   Measurement of f_r = _____

   Record the value of V_R and I at the resonant frequency.

   V_R = _____

   I = _____

2. *Lower half-power frequency, f_1:*

   f_1 and f_2 are the half-power or cutoff frequencies. At these frequencies, current and resistor voltages are 0.707 times their value at resonance. Calculate 0.707 of V_R you found in step 1 at resonance. _____

   Slowly reduce the source frequency and notice the lowering of V_R. (Note: It is especially important here that you maintain a constant input voltage.) When V_R reaches the value you calculated for f_1, read the frequency dial. This is the lower half-power frequency. Enter its value here and in Table 38–1.

   Experimental f_1 = _____

3. *Upper half-power frequency, f_2:*

   Using the same method, slowly *raise* the frequency above resonance. When V_R reaches the value you calculated for the half-power value, read f_2. Enter its value here and in the data table. _____

   Experimental f_2 = _____

4. *Bandwidth, BW:*

   Determine bandwidth from the relationship: $BW = f_2 - f_1$.

   Bandwidth = _____

Part 3—Circuit Measurements

1. Remove the inductor and measure the coil resistance. Add this value to the 560 Ω series resistance, or measure the coil and resistor together. This is the circuit resistance.

   Total circuit resistance = R_T = _____

   Replace the coil in the circuit if it has been removed.

2. Adjust the frequency to resonance by observing V_R with the oscilloscope.

3. Measure the following values with the oscilloscope. Remember to exchange the components as necessary to keep the part being measured at ground.

   V_T = _____ V_R = _____ V_C = _____

   V_L = _____ V_L and V_C combined = V_X = _____

   Enter the data in Table 38–1.

4. The following values are obtained by combining the measurements found above. Determine the following and enter the data in Table 38–1.

   Circuit current $(I) = \dfrac{V_R}{R} =$ _____

   Inductive reactance $(X_L) = \dfrac{V_L}{I} =$ _____

   Capacitive reactance $(X_C) = \dfrac{V_C}{I} =$ _____

   Net reactance $= \dfrac{V_X}{I}$ or $\dfrac{X_L - X_C}{I} =$ _____

   Circuit impedance $(Z) = \dfrac{V_T}{I} =$ _____

   Phase angle $(\theta) = \arctan \dfrac{X_L - X_C}{R}$ or $\arctan \dfrac{V_L - V_C}{V_R} =$ _____

   Circuit quality factor $(Q) = \dfrac{X_L}{R_T}$ or $\dfrac{V_L}{V_T} =$ _____

Table 38-1 Resonant Frequency Data: $C = 0.01\ \mu F$, $L = 100$ mH, $R = 1$ kΩ

Quantity	Calculated	Measured	Graph
f_r			
X_L			
X_C			
$(X_L - X_C)$			
Z			
I			
V_R			
V_L			
V_C			
$(V_L - V_C)$			
θ			
f_1			
f_2			
BW			
Q			

Part 4—Frequency Response Graph

If you take data for current over a range of frequencies and plot the values, you will have the frequency response for the circuit.

The circuit impedance at each frequency can be determined by the formula $\dfrac{V_T}{I}$.

A set of frequencies is suggested in Table 38-2, but you will need to find the exact resonant frequency yourself. Component tolerances are large enough to make a small variation in each team's data. Use the experimental value of f_r found previously, or check it again. Enter this frequency in the data cell reserved for it in Table 38-2.

Set the oscillator at the lowest frequency in the table, record current, calculate the impedance, and enter this data in Table 38-2. Obtain current and impedance for each frequency indicated.

Be certain to maintain a constant value of source voltage by monitoring the input on Channel 2. The actual value is not important, but maintaining it constant is!

Two sheets of semilog graph paper have been included. Use the first to plot current versus frequency. Use the second to plot impedance versus frequency. Consult with your instructor about plotting both curves on the same paper.

Part 5—Bandwidth, Q, and Half-Power Frequencies

1. If you have not already done so, plot the data for circuit current and impedance versus frequency from Table 38-2. Plot frequency on the horizontal axis—current or impedance on the vertical. Connect the points smoothly.

    The half-power frequencies, called f_1 and f_2, are the frequencies above and below resonance, where current is 0.707 (or square root of 2) times the current at resonance.

    The following method will let you determine f_1 and f_2 from the graph.

2. Calculate 0.707 of the maximum current. _____

3. Look at the left axis of your graph and place a small mark at the current value you calculated in question 2. Use a ruler or straight edge and draw a horizontal line at that value across both sides of the graph.

4. Plot a heavy dot where the line intersects the curve—on the left side and also the right. Mark the left intersection f_1 and the right intersection f_2.

5. Drop a line from f_1 and f_2 that intersects the frequency axis. The space between the two vertical lines represents the bandwidth.

6. Read the frequencies that correspond to f_1 and f_2.

    $f_1 = $ _____ $f_2 = $ _____

7. Since bandwidth is $f_2 - f_1$, subtract these two frequencies, and you have bandwidth obtained from your graph.

    $BW = $ _____ Hz

8. Circuit Q is $= \dfrac{f_r}{BW}$. Divide the resonant frequency by the bandwidth, and you have the circuit Q from the graph.

    $Q = \dfrac{f_r}{BW} = $ _____

9. Enter your data in Table 38-2.

Table 38-2 Data for Frequency Response Graph

Frequency Kilohertz	Current I	Impedance Z
2		
3		
4		
4.5		
f_r		
5.5		
6		
7		
8		
9		
10		

Part 6—The Effect of Component Values on Resonant Frequency

1. What effect does the capacitance size have on the resonant frequency? _____ If C becomes larger, what will happen to f_r? Will it increase, decrease, or remain the same?
   _____

   There are two ways to prove your answer. The first is to examine the equation for f_r:

   $$f_r = \frac{1}{2\pi\sqrt{LC}}$$

   Since C is in the denominator of the fraction, a *larger* C will generate the inverse or a *lower* resonant frequency.

2. The second method is to prove the answer experimentally. Remove the 0.01 µF and replace it with a 0.022 µF. Find the new resonant frequency. Record this value in Table 38-3.

   Did increasing the capacitance lower or raise the resonant frequency? _____

3. What will happen to frequency of resonance if the inductance becomes smaller? _____

   Use the same mathematical reasoning to obtain your answer. _____

4. Prove your answer experimentally. Replace the original 0.01 µF capacitor.

   Remove the 100 mH inductor and replace it with a smaller or 56 mH coil.

   Experimentally find the new f_r. _____

   Was your prediction in question 3 correct? _____

   Record your new frequency in Table 38-3.

5. What effect will a change in resistance have on the resonance? _____

   Mathematically, is R contained in the equation for determining f_r? _____

6. Confirm this with your circuit as well. Use the original 100 mH coil and 0.01 µF capacitor. Replace the resistance with a 270 Ω resistance.

   Find f_r for the circuit and record in the table. _____

   Did the resistance value affect the resonant frequency? _____

Table 38-3 The Effect of Component Size on Resonant Frequency

Replacement Component	f_r
0.022 µF	
56 mH	
270 Ω	

Frequency Response Graph   Circuit: _____

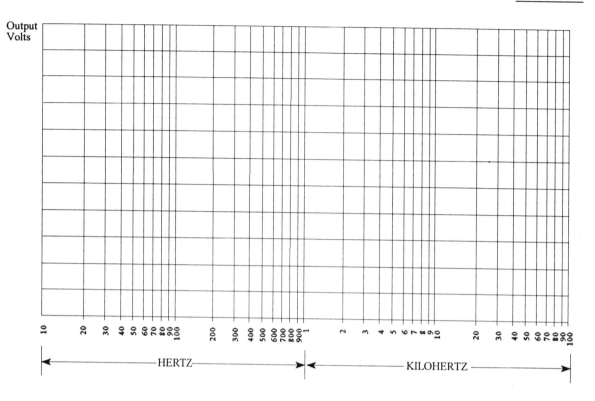

Resistance   _____   Resistance   _____
Capacitance   _____   Capacitance   _____
Filter Type   _____   Filter Type   _____
Cutoff Frequency _____   Cutoff Frequency _____

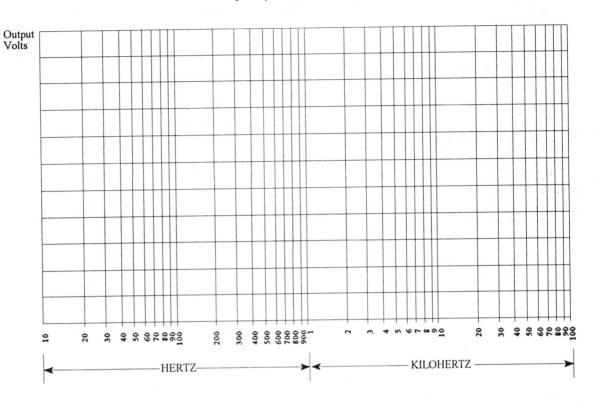

Frequency Response Graph   Circuit: _____

Resistance   _____   Resistance   _____
Capacitance   _____   Capacitance   _____
Filter Type   _____   Filter Type   _____
Cutoff Frequency _____   Cutoff Frequency _____

Lab Questions

1. If you want to raise the resonant frequency of the series circuit (Figure 38–1), would you use a larger or smaller capacitance? _____ Why would this work?

2. If you want to raise the resonant frequency, would you use a larger or smaller inductance? _____ Why?

3. Refer to your graph for impedance versus frequency. What is the value of circuit impedance at resonance? _____

    How does this compare with the total circuit resistance? _____

4. From your data in Table 38–2, did the change in series resistance affect the resonant frequency? _____

    If you did measure a change, what is the percent difference? _____

5. What effect does the coil resistance have on the bandwidth? _____

6. List the possible causes of experimental error in this project.

7. Describe an experiment that could demonstrate the effect of circuit resistance on bandwidth and f_1, f_2.

39 Parallel *RLC* Circuits

Objectives

1. Observe the currents and impedance of a parallel *RLC* circuit over a range of frequencies.
2. Obtain data and graph the resonant frequency response.
3. Read the graph to obtain resonant frequency, bandwidth, half-power points, and circuit Q.

Preparation

Part 1 looks at the currents above, below, and at resonance. You will measure the currents, observe the interaction of I_L and I_C, and determine phase angle.

Part 2 generates data for the frequency response graph.

The experimental procedures are similar to the measurement techniques used in the parallel *RL* experiment.

Materials

PC component assembler board
Single or dual trace oscilloscope
AC waveform generator
One 0.022 mH coil
One 0.01 µF capacitor
Resistors: ½ watt
    One 10 Ω, one 1 kΩ, one 10 kΩ
One 470 kΩ resistor

Procedure

Part 1—Currents and Phase Angle Measurements

1. Build the circuit of Figure 39-1.

$L = 22$ mH, $C = 0.01$ µF, $R_P = 1$ kΩ, and $R_S = 10$ Ω

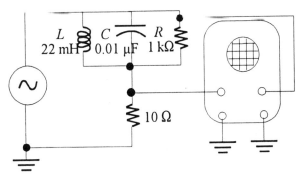

Figure 39-1

The oscilloscope is connected to monitor source voltage and measure the current sensing resistor.

2. Set source voltage to 10 V and frequency to 2 kHz.

3. Lift one lead of the capacitor and one lead of the resistor. This takes these two components out of the circuit.

4. Measure current flowing through the inductor. Enter in Table 39-1.

5. Measure coil current at each frequency indicated. Complete the table for I_L.

6. Measure capacitor current using the same method. Complete the data at each frequency.

7. Connect capacitor *and* coil. Open resistor. Measure current for *L and C combined* at each frequency.

8. Measure parallel resistor current. Complete the column for each frequency.

9. Find the phase angle at each frequency by using the formula:

$$\theta = \arctan\frac{I_C - I_L}{I_R}$$

Table 39-1 Data for Parallel *RLC* Circuits

Frequency	Measure I_C	Measure I_L	Measure $I_C - I_L$	Measure I_T	Calculate Phase Angle
2 kHz					
5 kHz					
10 kHz					
20 kHz					
40 kHz					

Part 2—Parallel Resonance

1. Calculate the resonant frequency of the tank circuit of Figure 39-1 using the formula:
$$f_r = \frac{1}{2\pi\sqrt{LC}}$$

2. Change the resistance values of the circuit. Use a 470 kΩ for the parallel resistance. Change the series resistor to 1 kΩ.

3. Set the signal generator to the resonant frequency calculated in step 1. Adjust input voltage to 4 V_{p-p}.

4. Slowly adjust the input frequency to either side of your calculated resonance. Observe the waveform across R_S. Since impedance is maximum at resonance, current will be minimum.

5. Adjust the scope sensitivity to make the display more readable when you are near resonance.

    When the pattern is minimum, this will indicate the resonant frequency. Read the frequency from the generator. Check frequency with the oscilloscope pattern. _____

6. Compare your experimental f_r with the calculated value.

7. Measure the total circuit current at resonance and enter in Table 39-2. _____

8. Lower the frequency 1000 Hz below resonance. Check the source voltage and adjust if necessary. Read circuit current. _____

9. Raise the frequency 1000 Hz above resonance. Check source voltage. Read circuit current. _____

10. Continue taking current readings at each frequency and complete the table.

11. Calculate impedance for each current and complete the impedance column. Impedance is obtained by dividing source voltage by circuit current.

12. Plot two curves on the semilog graph paper.

    The first graph is frequency versus current. The second graph is frequency versus impedance.

Table 39-2 Frequency Response Data

Frequency	Current	Impedance
1 kHz		
2 kHz		
4 kHz		
6 kHz		
8 kHz		
$f_r - 1000$		
f_r		
$f_r + 1000$		
12 kHz		
15 kHz		
20 kHz		
40 kHz		
70 kHz		
100 kHz		

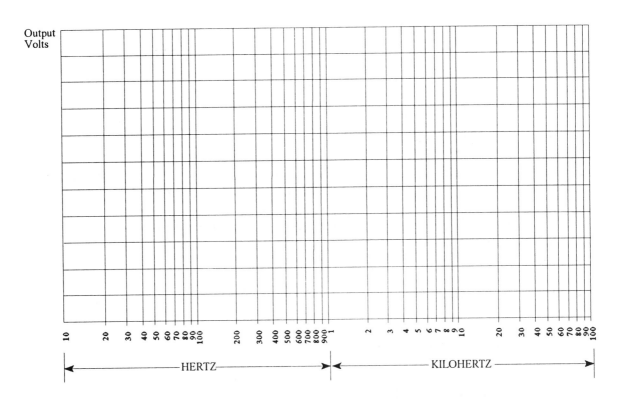

Lab Questions

Parallel RLC

1. From Table 39–1, which branch had the most current at low frequency? _____

2. Which branch had the most current at high frequency? _____

3. Notice the current columns for I_L and I_C. As I_C increased, I_L decreased. At some frequency, the current values crossed and were equal. What is the name of this frequency? _____

4. Notice the currents at 10 kHz. Why is the combination of I_L and I_C less than either I_L or I_C alone?

5. Did the parallel resistor current change over the range of frequencies? _____

6. Why did I_L and I_C change but I_R did not?

Resonance and the Frequency Response Graph

7. From Table 39–2, what was the circuit impedance at resonance? _____

   If I_L and I_C cancel at resonance and $I_X = 0$, what causes the impedance to be the measured value?

   List two sources of this impedance.

8. From the impedance graph of the circuit at resonance:

   Determine the lower rolloff frequency. _____

   Determine the upper rolloff frequency. _____

   Find the bandwidth. _____

   Calculate circuit Q from your graph data. _____

40 RC Filters

Objectives

1. Observe the frequency response of a low-pass and a high-pass filter circuit.
2. Plot the frequency response data on semilog graph paper.
3. Determine the cutoff frequencies from graphical data and compare the responses of different RC combinations.

Preparation

The first two circuits are low-pass *RC* filters. Since each has a different *RC* combination, the cutoff frequency will be different for each. By taking data over a range of frequencies and graphing the response curves, you can visually compare the filtering effects.

You can analyze the two high-pass filters in the same way. The graphs can be read to determine cutoff frequency for each combination.

Materials

PC assembly board
AC waveform generator
Dual or single trace oscilloscope
Capacitors:
    One 0.01 μF, one 0.047 μF
One 4.7 kΩ-1/2 watt resistor

Procedure

Part 1—Low-Pass Circuit

1. Build the circuit of Figure 40-1. This is a low-pass filter with the output taken across the capacitor. Resistance is 4.7 kΩ; the capacitor is 0.047 μF.

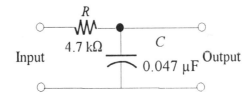

Figure 40-1 Low-Pass Filter Circuit

2. Set the input voltage to 2 V_{p-p} and frequency to 50 Hz. The pattern will blink at 50 Hz, but will become steady when the higher frequencies are used.

3. Measure the output voltage at 50 Hz. Since this is a low-pass filter, you can expect that the output may be about the same value as the input.

   Record the output in Table 40-1 for 0.047 μF.

4. Change frequency as indicated and measure output. Be sure to keep input constant.

5. Change the capacitor value to 0.01 μF. Keep resistance the same.

   Take measurement data for this new capacitor. Record in Table 40-1.

Table 40-1 Low-Pass Output Volts

Frequency	0.047 μF	0.01 μF
50 Hz		
100 Hz		
200 Hz		
500 Hz		
1000 Hz		
2000 Hz		
4000 Hz		
7000 Hz		
10 kHz		
20 kHz		
40 kHz		
70 kHz		
100 kHz		

Part 2—High-Pass Circuit

1. Change circuit to a high-pass filter arrangement (Figure 40-2).

$$R = 4.7 \text{ k}\Omega; \; C = 0.047 \; \mu F$$

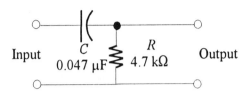

Figure 40-2 High-Pass Filter Circuit

2. Repeat measurements and record in Table 40-2.

3. Change capacitance to 0.01 μF and complete measurements.

Table 40-2 High-Pass Output Volts

Frequency	0.047 μF	0.01 μF
50 Hz		
100 Hz		
200 Hz		
500 Hz		
1000 Hz		
2000 Hz		
4000 Hz		
7000 Hz		
10 kHz		
20 kHz		
40 kHz		
70 kHz		
100 kHz		

Part 3—Graphing the Data

There are two sheets of semilog graph paper included with the project. The first is used to plot data for the two low-pass filters. The second is used to graph data for the two high-pass filters.

1. Label the vertical axis (the Y axis) with an appropriate scale of voltage.

2. Plot the response curves for each low-pass filter.

3. Identify each curve by labeling the circuit components at the start or end of each curve.

   Below the graph, fill in the values for resistance, capacitance, and filter type where indicated.

4. You can find the cutoff frequency by the following method.

   Calculate the output voltage equal to 0.707 of the input. Find this value on the left hand axis; read horizontally across until you intersect the curves. Read down from the intersection and obtain the frequency value on the horizontal axis.

   There are several names used for this frequency. Some of them are: cutoff, half-power, -3DB, and corner frequency.

   Record the cutoff frequency for each curve on the graph sheet where indicated.

5. Repeat the procedures for the high-pass data.

Frequency Response Graph   Circuit: _____

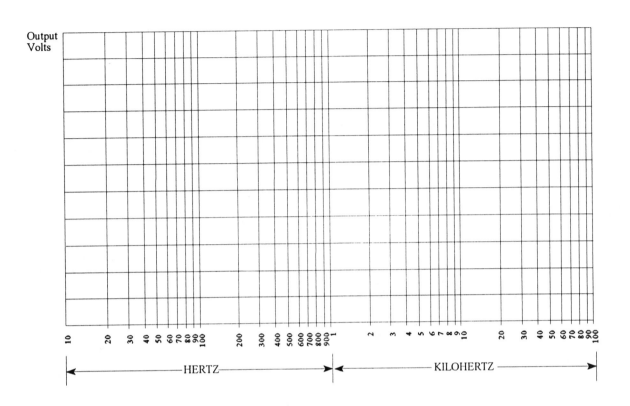

Resistance _____   Resistance _____

Capacitance _____   Capacitance _____

Filter Type _____   Filter Type _____

Cutoff Frequency _____ Cutoff Frequency _____

Frequency Response Graph   Circuit: _____

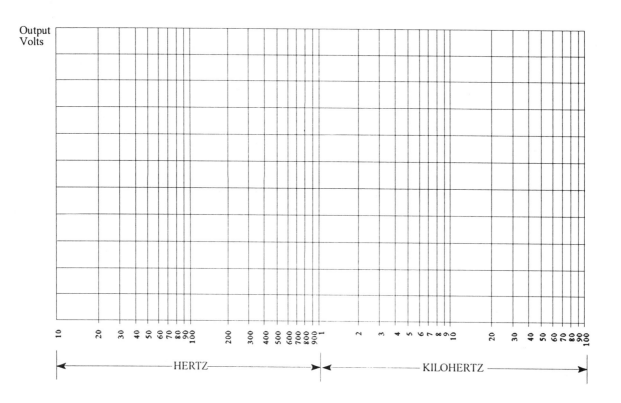

Resistance _____   Resistance _____
Capacitance _____   Capacitance _____
Filter Type _____   Filter Type _____
Cutoff Frequency _____ Cutoff Frequency _____

Lab Questions

Low-Pass Filter

1. From the data for your low-pass filter, which *RC* combination had the lowest cutoff frequency? _____

2. The cutoff frequency can be calculated from the equation: $f_C = \dfrac{1}{2\pi RC}$

   Calculate f_C for the 0.01 μF capacitor with the 4.7 kΩ resistance. _____

3. Calculate f_C for the 0.047 μF capacitor with the 4.7 kΩ resistance. _____

4. Do your calculations confirm your observations for the relationship of *RC* values to the cutoff frequency?

5. Two series *RC* circuits are being used as high-pass filters:

   Circuit A consists of 0.047 μF in series with a 3.3 kΩ resistance.

   Circuit B consists of 0.01 μF in series with a 3.3 kΩ resistance.

   Which circuit will have the *higher* cutoff frequency? _____

High-Pass Filter

6. From the data for your high-pass filter, which *RC* combination had the highest cutoff frequency? _____

7. The cutoff frequency (f_C) for the high pass filter can be calculated from the equation:

   $$f_C = \dfrac{1}{2\pi RC}$$

8. Calculate f_C for the 0.01 μF capacitor. _____ ____

9. Calculate f_C for the 0.047 μF capacitor. _____

   Do your calculations compare with your results from experimental data and graph? _____